A CRUISER'S GUIDE
TO
OCEAN WONDERS

Sabin Robbins

Edited by Will Robbins

Cover design by Byron Mucke - www.logicalrhino.com
Cover input by Rob Robbins.

To order copies (including autographed/personalized),
contact the author directly:

Sabin Robbins
3101 S. Ocean Blvd. Suite 510
Highland Beach, Fl. 33487-2510
U.S.A.
Phone: (561) 243-3122

CONTENTS

INTRODUCTION

Author at the lecture podium.

As a longtime National Geographic writer, zoo director, and lecturer, the most frequent question I am asked is, "Do you have a copy of your talk?" The answer is now in your hands.

Each illustrated chapter focuses on a sea subject that ranks highest in interest and enjoyment based on audience surveys. You'll learn—without getting wet—about the secret lives of whales, dolphins, seals, sea lions, sharks, and even pirates. And you'll find out the "facts" about the seas' most famous monsters and mysteries.

We live on a planet that is 70% covered with water—home to most of our global wildlife. I hope this book is a small step toward introducing you to the wonders of the ocean. Because in the end, we will conserve only what we love. We will love only what we understand. And, we will understand only what we learn and experience.

CHAPTER 1

THE SECRET LIVES OF WHALES

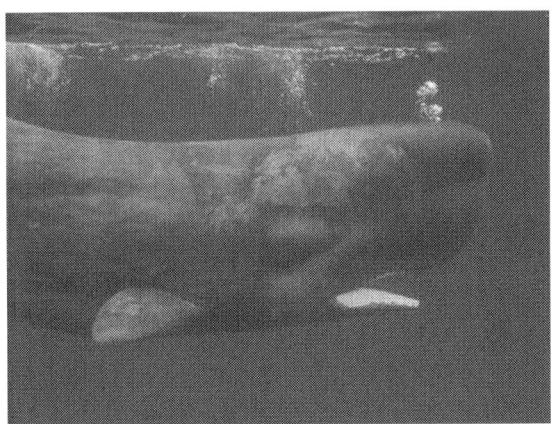

A denizen of the deep.

They move through a vast, dim, dark, watery world of their own. They are timeless and ancient, part of our common heritage and yet remote. They roam the ocean floor half-a-mile down, under the guidance of powers and senses we hardly understand. For we who rule the land, still know little of those that rule the deep. Their behavior still mystifies us. Their intricate sounds puzzle us. Their movements remain largely unknown. Their elaborate brains remain a riddle as do their lifestyles and relationships with each other. In every ocean, they swim and sing and socialize and travel as far as the sea is wide.

Like us on land, we know, however, that they are warm-blooded, must breathe to live, and just like us, give birth to babies that nurse on breast milk. They are our kindred spirits in the sea. They are the largest creatures ever to live on earth. They are, of course, the whales.

The story of man and whales goes back to the beginning of man. And our relationship with them is full of contradictions for we have feared and revered them, slaughtered and saved them.

Once called the "Leviathan" and "the great fish," they were feared as in the Bible's story of Jonah and the Whale. Because so little was known about them, it's not surprising that they were wrapped in mystery—and feared. When people first saw dead whales washed up on their beach that were 80-feet long and weighed 100 tons, they called them "demons of the deep." Early experts were sure that whales ate boats, and their bad breath caused brain damage.

Unfortunately, the theme of man's fear was soon followed by centuries of exploitation as killing whales made people rich because of the meat, oil, and other valuable parts. Rock carvings indicate men hunted whales with boats and harpoons 4,000 years ago. Europeans and Japanese were heavily involved in killing and processing whales 1,000 years ago.

Huge profits drove the industry. There was whale oil that lit the lamps of the world before electricity and lubricated the wheels of industry. The thin, flexible strips of whalebone made wonderful corset stays, hoops for skirts, umbrella ribs, buggy whips, clock springs, and fishing rods. They were even used as whips to discipline schoolchildren. Thus the child got "a whaling."

There was whale blubber used to make everything from soap and paint to margarine and dynamite. Ivory teeth were carved into buttons, necklaces, piano keys, and art objects. Whale skin was tanned for leather. Their skulls made farm walls, their flippers made glue, their tendons made tennis racquet strings, and their bones, blood, and guts were turned into pet food and fertilizer. Nothing was wasted—except the whale itself!

But if whales lost out by the millions, there was always danger for the whaler as well. Imagine being in a small, fragile boat in the middle of an icy, remote sea, and harpooning something 1,000 times heavier and stronger than you. A young Arthur Conan Doyle, creator of Sherlock Holmes, fell out of his whaleboat so often he was nicknamed "the Great Northern Diver."

On rare occasions, an enraged whale would attack the whaler. The most famous rogue whale was Mocha Dick, a 70-foot sperm whale that would be the model for Melville's *Moby-Dick*. For 30 years this giant battled the whalers that tried to kill him. When Mocha Dick was finally killed in 1859, they found 20 harpoons in his scarred hide from more than 100 battles in which at least 30 men had been killed and dozens of whaleboats destroyed.

By the mid-19th-century, 700 American whaling ships and 70,000 people were killing and processing whales. As Melville wrote, "The wealth of New England was harpooned and dragged up from the bottom of the sea."

It was a Norwegian whaling captain, Svend Foyn, who would lower the danger and increase the yield by building the first steam-powered and therefore speedy whale catcher. And he followed it up by inventing a cannon-fired harpoon that had a bomb in its tip. Later would come world-roaming factory ships and ship-based helicopters used as spotters. No whale anywhere in the world had a chance now. The pattern of killing was simple—hunt and kill the closest and easiest species until you can't find it anymore, then move on to the next species, and so on.

Whales were not just being used, they were being used up. Nearly two million whales were killed in just the 50 years prior to 1970. Even today in Japan, Norway, and Iceland, whale meat is sold in markets and served up in restaurants.

Too often, whales suffered long, agonizing deaths by harpoons. As one ex-whaler told me, "If whales could scream, whaling would have stopped years ago."

But popular opinion started turning in the '60s and '70s, especially in America, perhaps because so many whales can be seen along its shores. It began with Flipper and the performances of trained dolphins that caught the public interest, respect, and love for marine mammals. Books and television specials highlighted their fascinating behavior. The love affair with whales caught fire. Now it was whale lovers against whale killers as Greenpeace protesters put themselves between the harpoon cannons and the whales.

At the same time, millions watched whales from shores and boats making whale-watching a bigger business than whale-killing. Finally, the thousand-year-old war on whales is coming to an end. And happily, most whales are coming back in increasing numbers. In contrast to the past, we think about whales with equal parts of awe, wonder, reverence, and yes, guilt. These gentle giants have become an exalted, larger than life symbol of the environmental movement. And we humans have turned from predator to protector.

* * * * *

Hippo-like animals were likely land ancestors of whales.

If the story of humans and whales goes back a long time, the story of just the whale goes back a lot longer. Like other warm-blooded, air-breathing marine mammals like seals and sea lions, whales began as land-living mammals. Then 60 million years ago, they started to live in water. No one knows why. Perhaps there was more food in the water, or rising water forced them off the land.

DNA evidence indicates that the likely ancestor of whales was a relative of the hippo, whose legs evolved over millions of years into tail and flippers. Since whales are the marine mammals that have lived the longest in water, they are more fish-like than sea lions or seals. Yet they still carry reminders of their land ancestors. Their flippers have buried within them the bones of five fingers just like our hands. Very occasionally, whales have half-formed hind legs protruding from the rear of their bodies in a kind of genetic throw-back to the days when they had legs and walked on land.

All but a dozen of the some 80 species of whales have teeth and that includes all the dolphins and porpoises considered whales in all but size. Thanks to their teeth, scientists can estimate how old they are because their teeth have growth rings just like a tree.

Instead of teeth, a dozen species have baleen, a series of comb-like plates called *whalebone* that are flexible and fringed like a mop. Baleen is made of keratin, the same substance as our fingernails, and serves as a sieve to catch and filter the tiny fish and plankton they feed on.

Many whales have blotched skin because, like ships, they seem to attract lots of barnacles as well as crab-like lice.

All whales spout or breathe through their noses that evolutionarily have moved to the top of their heads to make breathing easier in the water. Experts can tell each species by the height and pattern of the blow since each species has differently-shaped nostrils.

The spout is not exhaled water. It's the warm, moist air exhaled from the whale's lungs that cools and condenses into moisture, like your breath does on a cold day. Having spent a lot of time with whales, I can tell you that their fish diet makes for some awful smelling breath close-up!

A whale's social life varies considerably. Some tend to be loners; others travel in groups called *pods*. Courtship often includes an increase in *breaching* or leaping out of the water. Courting can also include *lobtailing* or raising the tail above the water and then bringing it down with a mighty slap. Some whales will also check out what's going on by *spy hopping* or raising their head above the water.

Pregnancy is about a year. Newborns are generally born tail first. They are fully functional at birth, though will nurse for a year or more. Whale milk has the texture of canned milk but is ten times richer than cow's milk. It has a nutty not fishy taste so would be perfect on your strawberries at breakfast!

Calves will take piggyback rides on their moms. They reach maturity at six to 13 years, and, depending on the species, live 30 to possibly 200 years.

One of the most puzzling of all whale behaviors is their occasional determination to beach themselves and die either in shallow water or on the beach. And when people push them back out to safety, they return to the beach and sure death. In Argentina, 835 whales stranded and died at one time in 1946. Some years later, more than 1,000 whales at one time stranded and died along 20 miles of beach on Cape Cod.

Since circumstances and species vary, it's difficult to pinpoint a single cause of stranding. Over the years, many experts have given many explanations. Many scientists believe that the whales have gotten disorientated or lost because of a break-down in their navigation system. But that doesn't explain mass strandings. Some say their brains or inner ears are infested with parasitic worms that confuse their navigation. But the same worms are found in non-stranded whales. Some blame the super-powerful sonar used on Navy ships to detect enemy submarines, but many whales strand where there are no Navy ships.

Then there are the less persuasive theories. Earthquakes or storms trigger panic. It's the phases of the moon. Or a painful migraine headache causes the whale to throw itself onto the beach. And finally, whales are committing suicide to protest the whaling industry!

What may be truest of all is that we are no closer to solving the mystery than we were 2,000 years ago when Aristotle wrote, "It is not known why whales run themselves onto dry land. They seem placidly bent on self-destruction."

* * * * *

Cruisers often see high-jumping humpback whales.

Humpback whales (Megaptera novaeangliae/"Big-winged New Englander"), once 100,000 strong, were so successfully harpooned to make lubricants, detergents, and even lipstick and shoe polish, that they may only number 30,000 today. Fortunately, they are coming back under protection.

They can reach 50 feet and got their name because they expose a large part of their back that looks hump-like when they dive.

If you're bothered by ants at a picnic, pity the poor humpback that carries 1,000 pounds of barnacles on its body plus inch-long lice that infest its skin.

As the most acrobatic of all large whales, a humpback will often leap completely out of the water—and do it many times in a row. This is amazing as they can weigh 40 tons. When I was in Alaska a few years ago, a seaplane taking off was almost knocked out of the air by a high-jumping humpback. No one knows why they do it. Exuberance? A show of strength? To dislodge all those lice and parasites? Or perhaps it's part of the mating as they do it more during courtship.

The long front flippers that can extend a third of their body are unique to humpbacks, and gave them their Latin name meaning "Big-winged New Englander." Because of those long flippers, humpbacks are the only whales known to swim backwards.

Humpbacks have devised two unique feeding techniques. Lunging through schools of fish, they'll stun them with a slap of their flippers or tail, then gulp the fish down.

More remarkable, humpbacks make their own fish nets using bubbles. They begin by swimming beneath a school of fish, and blow out air to create a "net" of bubbles that surrounds the fish. Then with mouth open, they swim up for a quick meal. They even make different-size bubbles for different-size fish!

Of all the wondrous things humpbacks do, the most amazing is their singing ability. No other species makes more diverse sounds—clicks, grunts, groans, and whistles. Only the males sing, and mostly in their mating areas, so it seems to be part of their courtship. Observers say that the males that sing the best songs get the most girls so they call these whales "Engelbert Humperbacks." Their singing is so remarkable that when the space capsule, Voyager IV, rocketed beyond our solar system, it carried a greeting intended to convey the essence of earth cultures that included 60 human languages—and a humpback song.

Whale expert Roger Payne, who recorded a best-selling album called "Songs of the Humpback Whales," said that all the whales in a given group sing pretty much the same song, and then amazingly change that song every year. The whales don't just sing mechanically. They compose as they go along, incorporating new elements into their old songs. And the songs from one year to the next can be as different as Beethoven from the Beatles. No other animal except man does this, so it remains one of nature's great mysteries.

As author Peter Matthiessen wrote, "No word conveys the eeriness of the whale song, tuned by the ages to a purity beyond refining. It is a sound we should hear each morning to remind us of the dawn of the world."

* * * * *

California gray whales (Eschrichtius robustus) make the longest annual migration of any mammal on the planet—8,000 miles from Mexico to Alaska and back. That's like swimming to the moon and back over a lifetime.

They mate, give birth, and nurse their young during winter months in the lagoons of Baja California. By late-spring, the 20-foot-long babies join their 40-foot moms in pods up the Pacific coast to the fish-rich waters of the Arctic where they gorge on plankton, crabs, and shrimp before returning to Baja in late fall. They can average 60 miles a day.

Harpooners pushed gray whales to near-extinction.

Like humans, gray whales are right-handed or left-handed. A "right-hander" prefers to feed by scooping up food along the bottom using the right-side of its baleen. So that side is more worn down and scarred than the left-side.

California grays were called "Devilfish" by harpooners because they fiercely defended themselves and their young. But they were no match for harpooners who reduced a population of 25,000 to less than 100 by the late 1930s. One of the earliest Americans to fight for their survival was a well-known lover of women as well as whales. It was movie star Errol Flynn who financed the first-ever aerial photo survey that led to protection that has rebounded their population to a healthy 20,000.

* * * * *

Sperm whales (Physeter catodon), thanks to Melville's *Moby-Dick*, have become the symbol for all whaling. Their 60-foot length and 50-ton weight make them the largest of all toothed whales, and therefore the largest, large-prey predators that have ever lived. Larger than even Tyrannosaurus rex.

Sperm whales often battled whalers.

Long before Melville's *Moby-Dick*, sperm whales were hunted around the world because their huge, bulbous heads held up to 1,000 gallons of a fine, clear oil used in luxury candles and creams.

Sperm whale oil is even crucial in space age technology since it is the only known substance that works as a lubricant in the extreme cold temperatures of outer space. Said one scientist, "the Hubble space telescope is wheeling around the earth on spermaceti seeing six billion years into the past."

Sperm whales got their name because early whalers thought the oil in their heads was the whales' sperm. Admittedly, the head is an unusual location for sperm!

Even more valuable and something only sperm whales have is a waxy substance they secrete in their intestines called ambergris. It was highly sought after in perfume-making because of its astounding power to hold or "fix" the delicate scents in luxury perfumes. A 920-pound lump washed up in Australia in 1953 and sold for $320,000.

It's been said that ambergris was rubbed by Stradivari on his violins, and that was the secret ingredient that produced the violins remarkable sounds.

Herman Melville, by depicting his great white whale as an angry ghost of terror and destruction, demonized whales as effectively as others had demonized sharks and wolves. But Melville was just reflecting sailors' belief that sperm whales were killers bent on destroying men and ships. Although sperm whales never came close to living up to their evil reputation, Melville didn't write total fiction either. There were, in fact, instances where sperm whales turned on the seamen trying to kill them.

On the morning of November 20, 1820, the American whaler, *Essex*, was hunting whales in the Pacific. While the ship's three whaleboats were harpooning a pod of sperm whales, the crew back on the *Essex* watched in horror as a huge, 80-foot bull male surfaced, took aim, and came shooting at them, like a torpedo. A 70-ton torpedo! Cresting waves churned around its massive head. When it slammed into the *Essex*, there was a tremendous crash, and the ship shook as if she'd hit a rock.

Without hesitation, the enraged whale turned, and again attacked. This time it smashed in the hull's thick oak timbers. In 10 minutes, the 238-ton vessel capsized. The sailors had only a few minutes to get into the remaining whaleboats, and then return to salvage some supplies. Only eight of the 20-man crew survived after being lost at sea for three months. Never before had a whale attacked—much less sunk—a whale ship!

Like people, mother whales keep their babies close-by for feeding and protection. When mom is off feeding, other whales will baby-sit the young. In contrast to their *Moby-Dick* reputation, whales have demonstrated remarkably caring behavior—not only to their own kind—but to humans.

Recently off San Francisco, a whale became entangled in a dozen crab traps and rope. She was so weighted down by the thousand pounds of traps and hundreds of yards of rope wrapped around her that she was struggling to stay afloat to breathe.

A rescue team of divers spent an hour cutting off the ropes knowing that one slap from the 50-foot, 50-ton giant could kill them.

But instead, the whale floated passively, watching carefully as the divers slowly freed her. When the whale was free, she swam to each of the four divers who had saved her, and nuzzled each one as if to thank them. Then she swam away.

Said one diver, "The whale showed affection just like a dog that's happy to see you. I never felt threatened. It was the most incredibly beautiful experience of my life."

* * * * *

The Blue whale (Balaenoptera musculus) is the whale of all whales. Not only is it the biggest of all whales, but the largest animal that has ever lived on earth. Double the weight of the biggest dinosaur, it is as tall as a 10-story building, the size of a jetliner, and the weight of 130 cars or 3,000 people. Its heart is the size of a VW. A child could crawl through its largest arteries. Its tongue is as big as an elephant. Its 20-foot-long jaws hold enough water to fill-up a three-room house.

Yet the world's biggest animal eats some of the littlest food-inch-long shrimp and krill. It eats five tons of those crustaceans a day, which is like eating an elephant every day.

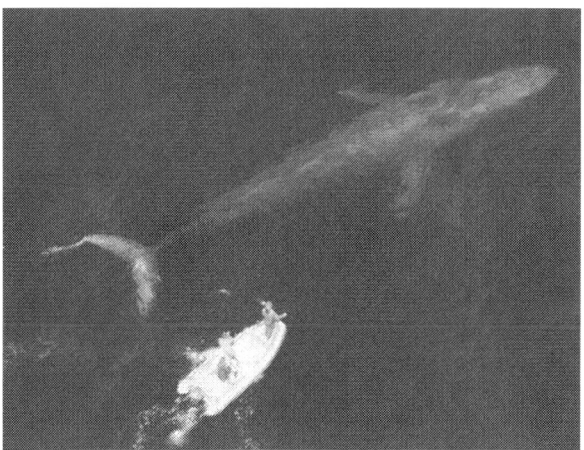

The planet's biggest animal.

At birth, the baby is 25-feet long, and drinks 50 gallons of its mother's milk every day. The fastest-growing baby in the world gains 250 pounds a day—or 10 pounds an hour. By the time it's weaned at eight months, it is 50-feet long and weighs 50,000 pounds.

There are probably fewer than 15,000 blue whales left on earth. Like many things about whales, the blue's migration routes and breeding sites are largely a mystery. Ironic that we know so little about the planet's biggest animal.

* * * * *

Years ago I got to know whales close-up. It was one of those magical and emotional experiences I will never forget, especially when these gentle giants and their calves came so close I could touch them. It was as if they were saying to me, "O.K. you once bad human, we are going to give you one more chance to be friends."

As humans, we have a choice about how we can interact with whales. The choice is simple: Hunt, kill, and use them. Or respect and protect them. Henry Beston must have been thinking about whales when he wrote in his book, *The Outermost House*:

> "We need a wiser and more mystical concept of animals. Civilized man—remote from universal nature and living by complicated artifice—surveys creatures through the glass of his knowledge and sees thereby a feather magnified and the whole image in distortion.
> "We patronize animals for their incompleteness, for their tragic fate of having taken form so far below us. And therein we err, and greatly err. For animals shall not be measured by man. In a world older and more complete than ours, they move finished and complete, gifted with extensions of the senses we have lost or never attained, living by voices we shall never hear. They are not brethren. They are not underlings. They are other nations, caught with ourselves in the net of life and time, fellow prisoners of the splendor and travail of the earth."

The author makes a whale of a friend.

* * * * *

FURTHER READING

The Book of Whales, Richard Ellis, Alfred A. Knopf, 1988.

Men and Whales, Richard Ellis, The Lyons Press, 1999.

Whales, Dolphins and Porpoises, National Geographic, 1995.

Whales and Dolphins, Vic Cox, Crescent Books, 1989.

Whales, Dolphins, Porpoises, Reader's Digest, 1997.

Whale Primer, Theodore Walker, Cabrillo Historical Assoc., 1979.

The Whalers, Time-Life Series, 1979.

World's Whales, Minasian & Balcomb & Foster, Smithsonian Press, 1984.

Whales and Dolphins, A. Cleave, Magna Books, Inc., 1993.

Whales, Dolphins and Porpoises of the World, Mary L. Baker, Doubleday, 1987.

The Spirit of the Whale, Jane Billinghurst, Voyageur Press, 2000.

Whale Watching, Nicky Leach, Ed., Discovery Insight Guide, 1999.

Whales and Other Sea Mammals, Time-Life Books, 1977.

In Search of Moby Dick, Tim Severin, Basic Books, 2000.

CHAPTER 2

AMAZING DOLPHINS AND PORPOISES

Four dolphins "perform" in the wild.

Thanks to movies and television, many people love them, though some people kill and eat them. Others play with them. Trainers teach them clever tricks. Psychologists probe their unfathomed minds. The military studies their super-sensitive guidance skills. A few scientists have even tried to communicate with them. And many view them with almost mystical awe.

Few observers can remain indifferent to these intelligent, playful, sensual, and intensely social creatures who seem as curious about us as we are about them. Welcome to the world of "Flipper,"–the dolphins and porpoises that belong to the same order, Cetacea, as whales.

Dolphins and porpoises are, in fact, just smaller-sized whales. And like whales—and humans—these marine mammals are warm-blooded, must surface to breathe air, bear young live, and nurse them from mammary glands.

The more than 40 species of dolphins and porpoises swim in rivers and oceans around the world. Because of confusion about the difference between dolphins and porpoises, the names are often used interchangeably.

Dolphin above and porpoise below.

In the picture above, a bottlenose dolphin or Flipper is above a harbor porpoise. The dolphin has a long nose or beak while the porpoise has a smaller, rounder look. Porpoises are typically shyer, and don't travel in large, leaping pods like dolphins.

Adding to the dolphin-porpoise confusion, there is also a brightly colored fish called dolphin, which is unrelated to the marine mammal we call dolphin. So many people thought restaurants were serving up Flipper that dolphin fish are now listed on menus as Dorado or Mahi-Mahi.

Like whales, dolphins and porpoises were once land animals that moved to water starting 60 million years ago. And like whales, they made specific adaptations for life in water over millions of years. Because water is 800 times denser than air, their bodies have become streamlined to reduce drag. Body hair and external ears have slowly disappeared to further reduce drag. Feet are fine on land, but flippers and tail are better for water. Yet X-rays still show the bones of five toes buried beneath their flippers and tail.

Dolphins and porpoises have also evolved thick blubber that keep them warm in cold water and provide buoyancy. Since water visibility can be poor, these marine mammals have developed acute hearing and communicate using clicks and whistles. Salt in ocean water can kill, so they have evolved kidneys and glands that remove excess salt through their urine and tears.

As a result of all these adaptations for water living, the whales and dolphins that have lived in water the longest, look the most like fish. So fish-like, in fact, that Catholics were once allowed to eat them on Fridays!

Dolphins are famous for being one of nature's most accomplished acrobats. So when you see them performing high, graceful leaps and somersaults, they are just doing what they have been doing naturally for millions of years.

How dolphins and porpoises sleep has baffled scientists for centuries. Aristotle claimed he heard dolphins snore. Some were sure that dolphins got out of the water and spent the night curled up on shore. Others were convinced they just took one big breath, swam to the bottom, and slept the night away. As it turned out, the truth is stranger than fiction.

Unlike humans who sleep because breathing is automatic, dolphins must control their breathing consciously by being awake to breathe or they'd drown. So they swim slowly at or just below the surface and take cat naps by going into a semiconscious state by switching off half their brain at a time. When one side of their brain is asleep, the other side is awake to control breathing—and watch out for predators.

Dolphins are sexually mature at six years, pregnancy is about a year, and babies nurse for almost a year. A mother dolphin will use another dolphin as a nurse maid to help raise her baby. The nurse maid is the only other dolphin allowed near the baby. Like humans, dolphins get cancer, heart disease, pneumonia, and ulcers.

Also like humans and few other mammals, dolphins enjoy social sex outside a restricted mating period. Some dolphins enjoy multi-partner sex, even kinky sex. They've been seen trying to make love to seals, turtles, and sharks!

Family groups typically stay together for life. So dolphins invest a lot of time raising their offspring. It could be said that for millions of years dolphins have been practicing the human adage that it takes a village to raise a child.

Dolphins will pack together in times of danger, and have been known to assist injured dolphins by pushing them to the surface to breathe.

Dolphins also work in teams to drive schools of fish to the surface to catch and eat them. Along coastlines, they will herd fish into shallow water, and then rush-drive them onto shore and eat them leisurely on the beach.

Dolphins have their own version of the telephone and e-mail. They operate their own sonar system by sending out high-frequency sounds from their rounded forehead that are so sophisticated they can see undersea creatures in X-ray fashion, even whether a shark's stomach is full or empty—and therefore whether their enemy might be feeding or not. Captive dolphins can tell when their female human trainers are pregnant before the trainers do!

At sea, dolphins recognize ships by the distinctive sounds they make so avoid boats that have been used to harass or hunt them. They also appear capable of long-distance communication by relaying messages among different groups.

Researchers work to unlock dolphin secrets.

Few marine mammals have been as intensely studied as dolphins. Scientists want to know how dolphins can go down 75 feet in a minute and dive as deep as 1,700 feet without decompression injury. The military wants to know why dolphin sonar is 16 times better than theirs. The Navy wants to train them to identify and retrieve objects for search-and-rescue missions.

Dr. John Lilly, in studies in the Bahamas and California, tried to learn their language of clicks and whistles so he could "talk" to them. He even tried to teach them English. He concluded that a dolphin's intelligence was between a dog and a chimpanzee. He was convinced their memory matched humans. New studies reveal that dolphins can recognize themselves in mirrors, even inspect their own bodies. The only other self-aware creatures are humans and some primates.

Few animals demonstrate their cleverness so often when working with humans. At a marine park in California, the trainers tried to get the dolphins to tidy up their pool by giving them a fish reward every time they brought back trash. One 12-year-old dolphin named "Mr. Spock" kept claiming fish after fish by retrieving soggy scraps of paper. Finally, the trainers realized that old Mr. Spock was keeping a pile of paper debris in a far corner and tearing off one piece at a time to get lots of fish. Mr. Spock was training the trainers!

Humans have always known that dolphins were special. Aristotle wrote 2,000 years ago that, "The dolphin is the only creature who loves man for his own sake."

There have been many stories about dolphins not only playing with humans—but rescuing them from danger and drowning. In parts of the world, dolphins even help humans catch fish. Fishermen set nets close to shore, then slap the water with sticks. The noise attracts the dolphins who swim toward shore driving schools of fish into the nets. The grateful fishermen always give some of their fish back to their dolphin helpers.

Some researchers claim the energy from dolphin sonar triggers the production of natural pain-killers (endorphins) that bolster the human immune system. Considerable research is being done on how dolphin-assisted therapy may help sick and handicapped humans. But those who oppose captive dolphin programs argue that the same results can come from using dogs and horses.

Fishermen in Vietnam have long given dolphins formal funerals that begin with burial in a special graveyard. Three years later, the bones are dug up, carefully washed, and re-assembled. Then pallbearers carry the remains in an elaborate hearse to the Temple of the Dolphins where fishermen pay their last respects by burning candles and incense, setting off fireworks, pounding drums, and playing songs of mourning.

Despite our attraction to dolphins, they are still wild animals and have been known to attack humans if they feel threatened. Ric O'Barry, the trainer of the six dolphins used in the "Flipper" television series, told me about one called "Patty" that was often bad-tempered and belligerent during filming. One day when Patty threatened an actor, Ric decided to teach Patty a lesson by thumping her back. With no immediate response, Patty continued swimming to the other side of the lagoon. She then turned slowly, sped up, and went for Ric like a rocket. The next thing Ric remembered was waking up in the hospital with a concussion.

In February, 2010 in Florida, a performing male orca (largest in the dolphin family) without warning attacked and drowned his trainer. At marine parks in the last 20 years, there have been at least six similar attacks, including a death at a Spanish zoo.

It was a reminder that dolphins are predators, just like sharks, and must kill to live. Male dolphins can be aggressive and heavily scarred from fighting. They've been known to gang up and kill smaller dolphins and porpoises. So we can misunderstand animal behavior when we project human feelings on them.

One of the most remarkable human-dolphin relationships began in 1888 when a dolphin began escorting ships across Pelorus Sound in New Zealand. The dolphin would do repeated jumps near the ships to attract attention. Then it would lead the ship safely through the shallow, dangerous channel. So good was this self-appointed pilot that "Pelorus Jack," as he was called, was given life-long protection by a special government order. Mark Twain and Rudyard Kipling witnessed Pelorus Jack in action from their ships.

Locals considered the dolphin a messenger from God because there was one and only one ship, the steamer *Penguin*, which he consistently refused to guide through the Sound.

Some travelers even refused to go on the ship, and in 1902 a crewman transferred off the *Penguin* because he said Pelorus Jack was sending a message of doom. Sure enough, the next week after the crewman left the ship, it ran aground, sank, and 75 passengers drowned.

One dolphin led ships through a dangerous channel.

Dolphin experts have a less mystical explanation for Pelorus Jack's message of doom. They say the ship was the only twin-screw vessel using the harbor, and the unusual and distinctive noise frightened off the dolphin. Whatever the reason, it is a fact that Pelorus Jack safely guided ships through the channel for 24 years until he disappeared in 1912.

Despite so many stories of dolphins going out of their way to help humans, Flipper is in trouble. Over the last few decades, millions of dolphins have been drowned in miles-long drift nets and huge purse seines used by commercial fishermen. Fortunately, some fishermen are now using special dolphin release nets. Nets are not the only killers of dolphins and porpoises. They've been herded into bays where they were trapped and killed for food. Others are poisoned by polluted waters, hit by speedboats, and have starved because over-fishing eliminated their food. No one knows how much noise and chemical pollution coastal dolphins can stand.

Our strong attraction to dolphins could be because in a world full of violence and fear, we feel safe and happy sharing our environment with a wild creature from the sea that not only accepts us, but, at times, seems to enjoy our company. As a Roman nature lover wrote 2,000 years ago, "There is no sea without dolphins."

* * * * *

A pinkish Amazon River dolphin goes airborne.

The boto or Amazon River dolphin (Inia geoffrensis) at 10 feet is the world's largest freshwater dolphin. Changing color from gray to pink as it matures has earned it another name, pink dolphin.

Though widespread through Amazonia, they also face threats from fishnets and pollution as well as dams and outboard motors.

The natives call them botos or "shape-shifters" because they believe these dolphins disguise themselves as men and come ashore during festivals to seduce girls. Rather than steal village maidens, botos use their needlelike teeth to eat piranhas as well as shrimp, crabs, eels, and turtles. They use echolocation to find fish hidden in the murky river waters.

* * * * *

Orcas are the largest dolphins.

Orcas (Orcinus orcas), largest members of the dolphin family, were described in the past by experts as "an enormous mass of flesh armed with savage teeth." These so-called "killer whales" have taken a bad rap for centuries because they do have big teeth which they use to catch their prey—seals, penguins, and even whales. Until recently, orcas were considered by most people as murderous maniacs that killed anything and everything.

Then in 1965, a young orca was trapped in a salmon net off Namu, British Columbia. "Namu" was exhibited in the Seattle

Aquarium where the Director swam with the 10-ton animal, hand-fed him salmon, and rode on the back of this so-called monster as if he was a big pool toy. The following year came "Shamu" to star at Sea World. The public now loved killer whales.

Orcas swim in every ocean from pole to pole so are the most widely distributed mammals on earth—and among the fastest at 30 m.p.h. They are easy to identify because of their striking black and white color and prominent dorsal fins that can be six-feet high. Alaskan natives believe that orcas take drowned souls to the undersea world.

Like humans, orcas nurse a year or more as infants, are sexually mature in their teens, experience menopause in their 40's, and live 80 years or more. Unlike humans, they have the biggest brains on earth—four times the size of humans.

Orcas are considered the ocean's most intelligent predator. Like wolves, they travel in packs up to 40. And like humans, they recognize foreigners by how they "speak." They also use their sonar to find and follow their prey. Orcas have even learned to follow fishing boats to steal from the nets. When fishermen are pulling in hooked tuna, orcas eat the fish as they're being hauled in.

Sometimes orcas will come right up to shore to catch a seal dinner. Yet while they hunt and eat seals in some oceans, they play with them in others. And though orcas clearly enjoy their all-meat dinners, no wild orca has ever killed a human. Mysteries abound.

As we share our cruise waters with these voyagers of a blue universe and enjoy the magic of their ways, we need to remember that if we hasten the disappearance of dolphins and porpoises, we diminish our world and our place in it.

Sharing the ocean's magic.

* * * * *

FURTHER READING

Whales, Dolphins and Porpoises, National Geographic, 1995.

Whales and Dolphins, Vic Cox, Crescent Books, 1989.

Whales, Dolphins, Porpoises, Reader's Digest, 1997.

Whales, Dolphins and Porpoises of the World, Mary L. Baker, Doubleday, 1987.

Whales and Dolphins, Mark Carwardine, Harper Collins, 1998.

Dolphins, Chris Catton, St. Martin's Press, 1995.

The Bottlenose Dolphin, John Reynolds, University of Florida Press, 2000.

Guide to Marine Mammals of the World, Audubon Society, 2002.

Whales and Dolphins, Carwardine, Hoyt, Fordyce, Gill, The Nature Company Guides. Weldon Owen Pty Limited, 1998.

Whales and Other Sea Mammals, Time-Life Series, 1977.

CHAPTER 3

SHARKS! MAGNIFICENT AND MISUNDERSTOOD

Sharks are the oceans' most demonized creatures.

Sharks (class Chondricthyes) have been called "death-dealing demons of the deep...mindless, killing machines...roving assassins...and insane, savage monsters." The word "shark" comes from the German word for "villain." Their very name describes danger or ruthlessness like card shark, pool shark, and loan shark. There's even a distinct shark vocabulary. Their teeth are always "razor-sharp." Their fins are always "slicing through the water." And the water is always "shark infested."

Many if not most people believe that sharks are the most dangerous creatures on earth because they kill more humans than any other animal; that they automatically go into a feeding frenzy when there's blood in the water; that their jaws are super-powerful; that they are specially attracted to bleeding humans; and that once they taste us, they search us out again and again.

It turns out that all those "facts" are false in the sense of having no scientific evidence to support such generalities. What is true is that sharks are the least known and most misunderstood of all sea creatures.

Sharks have fascinated and frightened people from the beginning. Scenes of sharks attacking humans were painted on cave walls by prehistoric artists thousands of years ago. Early seamen were sure the oceans were teeming with sharks just waiting to eat them. Sharks have starred as villains in books and movies from Melville and Hemingway to Peter Benchley's *Jaws*.

Polynesians once sacrificed their own people to appease the angry spirits of sharks. And some Pacific islanders still believe that sharks are the re-born souls of dead family members. So they build special shark temples where they use the skulls of ancestors to call up the spirits of sharks, who are then offered betel nuts and yams so they won't harm fishermen.

* * * * *

If human relationships with sharks are remarkable, so is their biology and behavior. Before there were dinosaurs, before there were trees, there were sharks. They've been swimming the seas for 400 million years.

Like other fish, and indeed all animals, sharks evolved in size, jaw, fin, and teeth. Since the business of life in the ocean is eat or be eaten, every creature is chased by another just a little bit bigger or faster. As a result, most sharks are streamlined, very sensitive to smell and movement and probably the fastest fish in the sea.

But sharks are fish with a difference. While fish have only one gill slit on either side, sharks have five to seven. While fish have bony skeletons, sharks have flexible, cartilage-like structures. Without a hard skeletal support, sharks can be crushed by their own weight when out of the water. Sharks also have no swim bladder that lets them float and rest like other fish, so they must swim or sink. Unlike bony fish that have smooth scales, sharks have tiny tooth-like denticles so abrasive that carpenters used their skin for sandpaper. Finally, unlike other fish that discharge sperm and eggs into the water, sharks mate directly and most, but not all, bear their young live like humans.

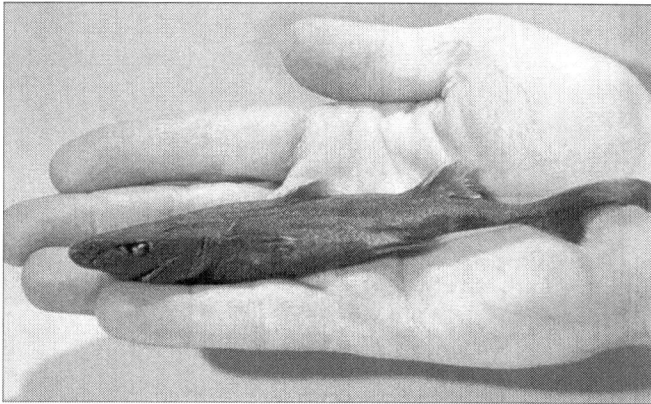

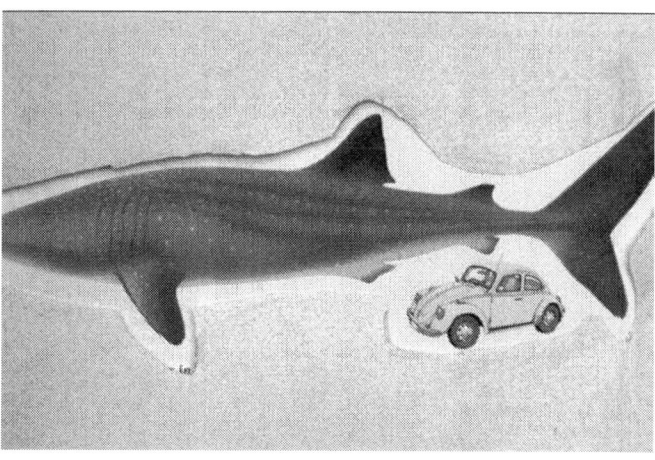

The dwarf shark is palm-size while the whale shark can be 60-feet long.

There is no "typical" shark. The 400 species can be as small as the six-inch dwarf shark and as big as the 60-foot whale shark. Some glow in the dark while others can change colors as fast as chameleons. Most are gray, but some are as brightly colored as goldfish. Some have stripes. Some have spots. Some spit and grunt, while others bark like dogs or croak like frogs. Some have bodies like eels, saws for noses, hammer-like heads, and faces like bulldogs.

Sharks swim in every ocean from the Arctic to the tropics. Others live in rivers and lakes far from the nearest salt water. Some swim in shallow waters while others cruise the deepest, darkest depths. Some are loners while others travel in schools.

Depending on species, sharks live 25 to 70 years. Like trees, they produce annual growth rings in their vertebrae that vary from one-inch to a foot a year.

Few creatures are as sensitive to sounds, smells, and movements. Thanks to electro-receptors in their heads, sharks can pick up the tiniest muscle movements of prey as far as a mile away. To a hungry shark, a fish in distress is a dinner bell. Their smell is so good that they can detect a drop of blood a quarter-mile away.

Since different sharks eat different foods, they have different teeth. Some have flat teeth for crushing shellfish while others have broad, serrated teeth for cutting apart larger prey like seals. Though some teeth can cut paper like a razor blade, claims that sharks have super jaw power are exaggerated. Tests indicate that Eskimos have twice the jaw power of sharks. Pit bulls even more.

Also exaggerated is that sharks eat anything and everything. Shark stomachs have supposedly included everything from a cuckoo clock and a keg of nails to a Frenchman in a full suit of armor. But when the National Marine Fisheries Laboratory examined hundreds of sharks, not one stomach had anything more exotic than a tin can or plastic bottle.

According to books and movies, sharks roam the world hunting people because they crave human flesh. Since sharks have been eating 400 million years before humans even evolved, one wonders how they survived all those years without us!

Of course, sharks do occasionally sample humans, but they much prefer the fattier flesh of fish, seals, and sea lions. Certainly sharks bite people—as dogs do and a lot more often— but it's usually because sharks mistake us for seals or turtles.

The "feeding frenzies" that are the horror highlight of movies are triggered by throwing buckets of fish and blood in the water to induce a non-typical behavior that makes for great action, but bad biology.

The rare feeding frenzies in nature are usually over dead carcasses like whales. Combine lots of food and lots of animals in a small space and you get competition. You can easily induce feeding frenzies in birds—or humans at late-night buffets on cruise ships!

But if shark feeding frenzies are rare, they still happen. During World War II in the Pacific, the battle cruiser, *Indianapolis*, was torpedoed by a Japanese submarine. The ship sank in 12 minutes, pitching 900 crewmen into the water. Within hours, hundreds of sharks circled. During the four days awaiting rescue, 400 seamen drowned and 200 were killed by sharks. It was the worst naval disaster at sea in American history.

Depending on the species, sharks have three different reproductive techniques. Most give birth to live young from internal eggs. Others package their eggs in sacks deposited on the ocean floor to await hatching. And some sharks give birth just like humans—with placenta, embryo, and umbilical cord. Pregnancy varies from seven to 22 months depending on the species.

At birth, baby sharks are fully independent—and have a full set of teeth. Even unborn young can be aggressive. When a veterinarian performed a C-section on a pregnant sand shark, the unborn dashed around—and bit him!

Sharks grow and mature slowly, producing far fewer offspring than other fish. As a result, their populations are easily threatened by overfishing or slaughter.

Only one mating of a large shark has ever been observed so what is known about their sex life is based on the behavior of smaller species in captivity. As far as migration, some sharks do and some don't. And some species do in some areas and not in others. No one really knows.

If what sharks do is hardly known, why they do things is mostly a mystery. Some species, like the great white, are more aggressive than others yet sometimes they bite and sometimes they don't. Sometimes they feed on a full stomach and don't on an empty stomach. Sometimes the same species behave differently in different places. The only thing that is predictable is their unpredictability. Since most animals are predictable, the idea that sharks are unpredictable is probably more a reflection of how little we know about them.

Concluded shark expert, Dr. Samuel Gruber, "I can train a lemon shark faster and more reliably than a rat, rabbit, or cat. So if they are smart, then sharks are smart."

* * * * *

Spiny dogfish (Squalus acanthias) are the most prolific sharks in the sea. One-fourth of all shark species are in the dogfish family. They average four-feet long, and are the Methuselahs of sharks living 70 or more years. While sexual maturity and pregnancy vary among species, female dogfish don't mature until 19 or 20, and pregnancy lasts two years. That's longer than whales, and almost as long as elephants. No one knows why.

Much of what we know about shark biology comes from scientists studying dogfish because they are abundant, easy to catch, and do well in captivity.

During one season off Massachusetts, commercial fishermen caught 27-million spiny dogfish. In Nova Scotia when one long- line of 700 hooks was pulled up, 690 hooks held spiny dogfish. Some fishermen hate dogfish because they tangle their nets trying to eat other trapped fish. But in Europe, fishermen catch and sell dogfish shark as "rock salmon" or "flake" used for "fish and chips" meals.

* * * * *

Tiger sharks are not picky eaters!

Tiger sharks (Galeocerdo cuvieri) at 20 feet and 2,000 pounds are one of the largest and most dominant of reef predators. They get their name from the dark stripes that cross their backs.

Tiger sharks have been called "swimming garbage cans" because they do seem to eat anything that moves—from fish and turtles to seals, dolphins, and even other sharks.

They also seem to eat a lot of things that don't move. Their stomach contents have included license plates, boat cushions, and a roll of tar paper. One tiger ate an unopened can of salmon, its favorite fish food. So if the label was still on the can, maybe sharks can read!

Another tiger shark was so infamous it had a name. "Shanghai Bill" allegedly ate dozens of sailors in the Caribbean before swallowing a big, shaggy sheepdog. The dog's hair apparently caught in the shark's teeth and choked it to death. So that may be the world's first shaggy dog story!

* * * * *

The bull shark (Carcharhinus leucas) is considered the most aggressive to humans after the great white shark.

At up to 11 feet and 400 pounds, they have very large teeth, a massive jaw, and eat turtles, fish, dolphins, and even young hippos.

Their danger lies not so much in their belligerent behavior and big teeth, but that they are at home in every water—salty ocean or freshwater lake. They swim up the Mississippi, Ganges and Zambezi rivers as well as the Chesapeake Bay, and have been seen 2,000 miles up the Amazon. They also like shallow water—where human swimmers are.

* * * * *

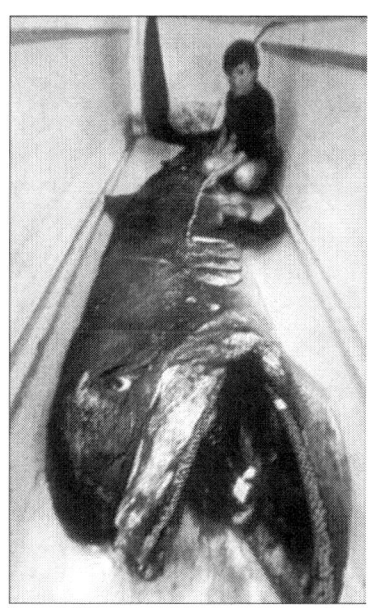

The mammoth megamouth was unknown until 1976.

The megamouth shark (Megachasma pelagios) remained undiscovered in the vast, deep Pacific Ocean until a Navy ship off Hawaii hauled in its parachute sea anchor in 1976. Tangled in the parachute was a never-before-seen species of shark that was 14-feet long and weighed 1,653 pounds. Not surprisingly when they saw its six-foot-wide mouth, it was named megamouth.

Since then more than a dozen megamouths have been caught or washed ashore along various Pacific shores from California to Japan.

Little is known about this newly discovered species. Based on its stomach contents, it is a filter feeder of plankton and small fish. Luminescent tissue inside its mouth indicates that it may swim with mouth open to lure in plankton and deep-sea shrimp with the faint light.

* * * * *

The friendly whale shark is the planet's biggest fish.

The whale shark (Rhincodon typus) is not only the largest of all sharks, but the largest fish in the ocean.

Long a creature of myth, the 40-foot, 13-ton behemoth only became a scientific reality in 1828 when one was caught off Cape Town, South Africa.

Like the megamouth shark, it uses its six-foot-wide mouth to suck up plankton and tiny fish on or below the surface. Its coloration—unique among sharks—is an intricate pattern of yellow or white dots and stripes in a rectangular grid pattern. You could play checkers on its broad back.

Whale sharks travel the tropical seas of the Atlantic, Pacific, and Indian Oceans. They have been seen as far north as Massachusetts.

So mild-mannered are whale sharks that divers can ride on their backs. But when a seamen foolishly harpooned one, the docile creature instantly changed into a fury-fueled mountain of steel muscles. Then in a cascade of water, the giant dove into the depths. As the harpoon's thick line flew through the air, it snagged on a raft and snapped like a piece of string. Seconds later the broken-off harpoon shaft popped to the surface. The whale shark was never seen again.

* * * * *

Great white sharks can attack by air.

There is no animal we fear more—or no less about—than the great white shark. Although they have been around for millions of years, they have never been seen mating or giving birth in the wild. The great white shark (Carcharodon carcharias) got its name for its size and white belly. And got its fame from Peter Benchley's 1974 novel, *Jaws*, and movie.

Centuries before *Jaws*, great whites terrified seamen who believed they leapt out of the ocean and pulled sailors out of the tall rigging as they sailed by.

It so happens that just a decade or so ago, great whites in South Africa were seen for the first time leaping completely out-of-the-water to attack seals swimming on the surface. They soon were dubbed "Air Jaws," and have repeated those airborne assaults in other parts of the world.

At 30-feet and 4,000 pounds (the weight of a rhinoceros), great white sharks are the world's largest predatory fish. An ambush hunter, they hide, wait, and strike. Like a torpedo with teeth, great whites are armed with rows of razor-sharp, three-inch teeth in a huge mouth. They are also the only sharks that lift their heads out of the water to pinpoint prey on the surface.

One expert called the great white shark, "a total carnivore...the last free predator of people...the most frightening killing-machine on earth."

* * * * *

No other shark behavior gets more publicity than shark attacks on people. The stories go back hundreds of years when the first humans went in the water. But in fact, only 27 of the 400 shark species have ever attacked humans. Worldwide in an average year, sharks bite fewer than 100 swimmers and kill 15. In contrast, 1,600 New Yorkers a year are bitten—by other New Yorkers!

Every year 40,000 Americans die in cars, but we still drive. Every year, 800 die on bikes, but we still bike. And every year, 150 people are killed by falling coconuts. So it's 10-times more dangerous to stroll a palm-shaded beach than swim in "shark-infested" waters!

Many more people die falling out-of-bed or getting kicked by a donkey than from sharks! Compared to the 15 humans killed by sharks every year, humans kill 70 million sharks by intention, accident, for sport or money. So the real story is not "shark bites man," but "man bites shark" since for every dead human, we kill seven million sharks.

Yet fear of sharks has triggered a centuries-old search for shark repellants. Researchers have tried chemicals, bubble screens, nets, electric and sonic barriers. Few, if any, really work. During World War II, an American intelligence officer cooked up a mix of ingredients that kept sharks away from mines but not humans. That officer would later become a cooking legend—Julia Child.

Recent studies suggest that sharks are not out for a human meal so much as defending their territory or mistaking humans, especially surfboarders, for their normal diet of seals and turtles. Only one-in-five attacks involve more than one bite as sharks much prefer fattier fish, seals, and sea lions. So sharks are typically man-biters not man-eaters.

Human fear, ignorance, and greed have made sharks the object of search-and-destroy missions around the world. So we've slaughtered them for their leather, their flesh, their fins, and for our "protection." Just in the last decade, shark populations have plummeted 80%.

Sharks are worth saving because they are a vital part of the delicate balance of nature. For example, when sharks were slaughtered in Australia, there was an explosion of octopus, their favorite food. So as sharks became few, octopuses became many—and proceeded to eat the lobsters that supported fishermen.

Sharks are also valuable to humans because their cartilage has been used to make artificial skin for burn victims, their corneas used in human transplants, and their liver oil promotes white blood cell production and is an active ingredient in Preparation H because it shrinks hemorrhoids.

Hopefully, the most dangerous creature on land and the most feared creature in the sea will learn to live together. In fact, that day may have arrived.

According to an Australian newspaper story, Andrea Lyons found her pet shark, Archie, so loveable that she married him. Said the beaming bride, "Everybody says Archie is dangerous and that someday he's going to eat me alive. But compared to men I've met lately, he's an absolute angel. He's gentle and he kisses nice and he doesn't leave his dirty underwear lying all over the house. As far as I'm concerned, he's one helluva catch."

Humans and sharks have begun to live together!

* * * * *

FURTHER READING

The Shark Almanac, Thomas Allen, The Lyons Press, 2003.

The Book of Sharks, Richard Ellis, Alfred A. Knopf, 1989.

Sharks in Question, Springer & Gold, Smithsonian Press, 1989.

Sharks, Doug Perrine, Motorbooks Intl., 1995.

Sharks: Myth and Reality, Cafiero & Jahoda, Thomasson-Grant, 1994.

Sharks!, J. Rotman, Ipso Facto Publisher, 1999.

Sharks: Silent Hunters of the Deep, Reader's Digest, 1986.

The Secret Life of Sharks, Peter Klimley, Simon & Schuster, 2003.

Sharks of the World, J.V. Compagno, 1984.

Great White Shark, Richard Ellis, HarperCollins, 1991.

CHAPTER 4

MONSTERS OF THE DEEP

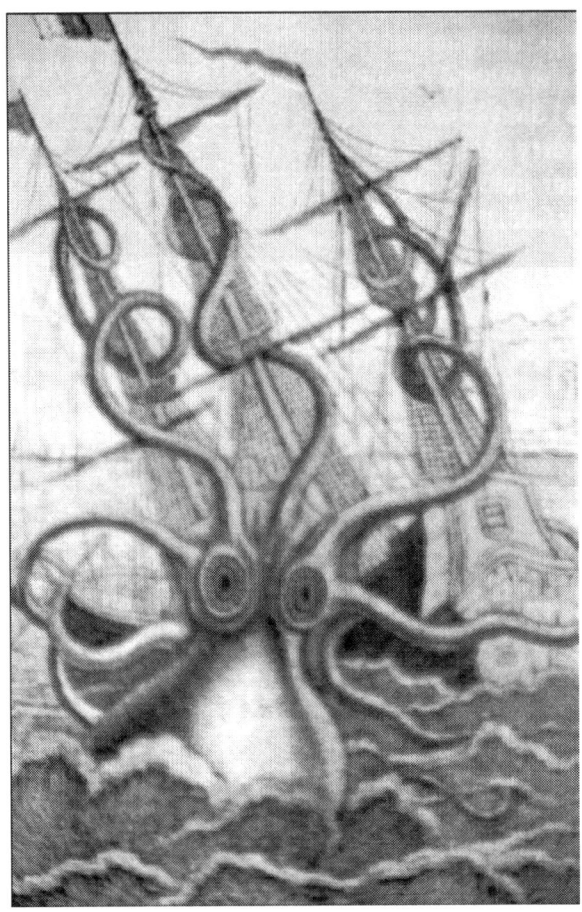

A tentacled sea monster attacks a ship.

According to a recent newspaper poll, half their readers believe in ghosts, a third in UFO's (Unidentified Flying Objects), and five percent believe that Elvis Presley is still alive!

Like ghosts and witches, spirit rappings and séances, sea monsters have been around for centuries. Thousands of eye-witnesses, drawings, and photographs have allegedly proven their existence. After all, sea monsters are as old as Jonah's whale and as recent as Jaws. Seamen have provided detailed accounts of sea monsters that ate sailors and swallowed ships. In earlier days, the vast oceans were largely unexplored. So mapmakers filled in the blanks with fanciful beasts and wrote, "Here Be Monsters."

As sociobiologist, E.O. Wilson, wrote, "Monsters provide a sweet sense of horror. The shivery fascination with monsters and creeping forms delight us today, even in the sterile hearts of the cities, because it could see you through to the next morning."

So an ocean without sea monsters is like sleep without dreams—or should I say nightmares.

* * * * *

Some ship-sized sea serpents spit seawater.

Sea serpents were documented in zoology books up through the 19th-century. They were described as snake-like, 50 to 100-feet long, and some had long hair and flaming eyes. They lived in caves deep in the ocean, and swam out at night to eat pigs, sheep, and, of course, sailors. As proof of their existence, alleged parts of them often washed ashore.

Danish missionary, Hans Egede, who later became the Bishop of Greenland, gave this first-hand report in 1734: "A most dreadful monster showed itself upon the surface. It was so huge that coming out of the water its head reached as high as our mast. Its body was as big around as our ship, four times as long, and covered with shells. It had a long, pointed nose that spit water."

In 1808, a decomposed serpent washed up in Scotland. Farmer John Peace said it was 55-feet long, had a long neck, and six toes on each hairy paw. But when samples were studied by a doctor naturalist, he said it was probably a basking shark, which are very real, the size of a bus, and a common sight off Scotland.

Americans were not to be left behind when it came to seeing sea serpents. In 1638 off Cape Ann, Massachusetts, a 60-foot sea monster came out of the water and coiled itself on a rock. Observers said its huge, snake-like head was bluish except for a black circle around the eyes. When it finally swam away, it moved with tremendous speed.

But the Cape Ann serpent was pint-sized compared to the Gloucester Monster of 1817, the biggest ever seen. During that fateful summer, hundreds of people saw it. Eye-witnesses said it was 100-feet long with a head as big as a horse's. Moving like a giant caterpillar, it swam around the harbor several times, and came so close to shore that one brave soul tried to harpoon it.

A month later, two boys playing on the beach found a three-foot snake with humps on its back. After dissection, scientists said it was a baby sea serpent. They concluded that the

Gloucester Monster had come close to shore so that it could lay its eggs and this was its baby.

The Gloucester Monster was seen by hundreds.

Later, however, a French expert examined the creature, and declared it was only a deformed specimen of the common black snake.

Sightings of sea monsters were so commonplace for the next 75 years that Dr. Antoon Oudemans, the renowned Director of Holland's Royal Zoo, recognized sea serpents as scientific fact in 1892. In his encyclopedic book, *The Great Sea Serpent*, he described in detail 162 confirmed serpent sightings. He concluded that these monsters did, in fact, reach 100 feet, lived in caves, and were mammals not reptiles.

Just since 1990, there have been more than 20 eye-witness reports of sea serpents swimming off Canada's Vancouver Island. All agree that Cadborosaurus, named for British Columbia's Cadboro Bay and nicknamed "Caddy," was over 50 feet and had three-to-five humps along its back. Caddy's head was as big as a camel's, and its long neck was covered with brown hair.

As reported in The New York Times in 1992, "The two Canadian scientists studying these reports believe that these large sea serpents may comprise a distinct vertebrate species related to the Loch Ness Monster."

Could "Caddy" be a 50-foot oarfish?

The real-life ribbon or oarfish could be mistaken for a sea serpent. As the world's longest bony fish at up to 50 feet, the eel-like oarfish has a flattened body like a snake and a bright red, spike-like crest or fin along the top of its body.

Swimming on the surface with its head poking out and red crest cutting the water could suggest a sea serpent. But hard-to-believe it inspired the 100-foot Gloucester Monster of 1817.

* * * * *

The most popular and romantic of sea monsters are mermaids—and mermen, vital to producing mermaids.

Like sea serpents, stories of half-human, half-fish creatures go back to ancient times. Mermaids allegedly loved music, lived long lives, and had magical powers. The more vicious mermen had mouths full of fangs—and attacked ships.

Mermaid with her merman.

Experts warned that mermaids enticed sailors with their melodious songs, put them to sleep, then pounced on them and tore them to bits.

Columbus saw three mermaids on his discovery of America, and wrote in his log "But they are not as beautiful as they are painted."

In contrast, when Henry Hudson discovered New York's Hudson River in 1608, he described a mermaid close to his ship as, "beautiful, her back and breasts were like a woman, her skin very white, and she had long, black hair...I threw a fishhook to see if she would bite, but she dove and disappeared for good."

As late as the 19th-century, serious natural history books included detailed descriptions and drawings of mermaids based on hundreds of sightings—and even a few captures!

The author of *The Natural History of East Indies* wrote that a five-foot-long mermaid caught off Borneo had lived four days in a barrel of water. "From time to time, it uttered little cries like a mouse. Though offered small fish and crabs, it would not eat."

In fact, scientific theory at one time supported the existence of mermaids since it was long believed that every creature on land had a counterpart in the sea. Thus, there were horses and sea horses, dogs and dogfish, cats and catfish, lions and sea lions so why not maidens and mermaids.

Englishman Robert Hawker pulled off an elaborate hoax in 1825 by stripping naked, wrapping his lower body in an oilskin tail, and swimming to a rock off the Cornwall coast. He spent several nights on the rock combing seaweed from his hair and singing to large crowds. For a finale, Hawker sang a mighty rendition of "God Save the King," and then dove off the rock and out-of-sight.

The 19th-century was the heyday for exhibiting mermaids made of everything from skeletons and mummies to fish parts glued together to look like mermaids. Of course, P.T. Barnum had to have his own mermaid—and so he did.

Barnum heard about a mermaid that a Boston sea captain had bought in India and wanted to sell. Barnum knew he had found a blockbuster attraction for his newly-opened American Museum in New York City. But simply buying and displaying the mermaid wasn't the Barnum style. He had to concoct his own story about how his mermaid had been caught in Fiji, preserved in China, and was now in the possession of a Dr. Griffin, who was taking it back to London's Natural History Museum. Dr. Griffin had kindly consented to let Barnum exhibit it for a week in 1842. Of course, the "Feejee Mermaid" stayed much, much longer—and raked in thousands of dollars a week.

Barnum's mermaid was no beauty!

When Barnum later wrote his autobiography, he gleefully revealed his hoax, admitting that Dr. Griffin was actually his accomplice, Levi Lyman. Barnum described his mermaid as, "an ugly, dried-up, black-looking specimen that looked like it had died in great agony." He guessed it was the result of someone surgically connecting a fishtail to a monkey's torso and head.

Most experts believe that a friendly marine mammal, the manatee, likely inspired the mermaid stories. But whether myth or manatee, mermaids are as popular today as ever. The statue of Hans Christian Anderson's Little Mermaid has always been Denmark's most popular tourist attraction, and the Danish Tourist Board claims it is also the world's most photographed statue.

* * * * *

The giant squid is a very real sea monster.

If the mermaid is the most loved sea monster, the most feared sea monster is likely Architeuthis, the giant squid. It is the planet's largest, living, invertebrate predator weighing over 2,000 pounds and reaching a length of 50 feet. It lives in a place so cold, dark, and deep that the elusive creature was only seen alive and captured on camera 3,000-feet deep off the coast of Japan in 2004.

In 1861, the crew of a French ship in the Canary Islands harpooned a giant squid but could only get its tail aboard. They estimated that the body was 18-feet long with six-foot tentacles for a total length of 24 feet. Soon after that, a story went round that several French ships were attacked and sunk by a colossal squid.

Those stories likely inspired Jules Verne because nine years later he wrote perhaps the most famous of all sea monster stories, *20,000 Leagues Under the Sea*. Verne's giant squid that attacked the submarine *Nautilus* was 25-feet long with blue-green eyes and a beak full of razor-sharp teeth. Its eight tentacles were armed with 250 suction cups. It turns out that giant squid not only exist but are even bigger than Verne's monster.

Three years after Verne's book, fishermen in Newfoundland brought home a dead giant squid that was quickly bought for $10 by the right Reverend Moses Harvey. As Reverend Harvey recalled, "I remember how I stood on the shore gazing on the dead giant and relishing how I would astonish the wise men, confound the naturalists, and startle the world."

The first giant squid goes on public display.

Sure enough, Harvey startled sightseers by hauling the beast home where he draped the head and arms over his bathtub. Then he proudly invited in a steady stream of goggle-eyed visitors. It's not known what Mrs. Harvey thought about a 32-foot dead squid decorating her home!

So a sea monster believed by many to be a myth proved its existence by washing up on beaches all over Newfoundland from 1871 to 1881—and off-and-on around the world ever since.

In 1991, popular writer, Peter Benchley, did for giant squid what he had done for great white sharks 16 years before in *Jaws*. He turned the giant squid into a man-eating, ship-grabbing monster in his best-seller, *Beast*. Here's how Benchley described his giant squid:

"It hovered in the ink-dark water, waiting... Its eight sinuous arms floated on the current. Its two long tentacles were coiled tight against its body. When it was threatened or in a frenzy of a kill, the tentacles would spring forward, like tooth-studded whips. It existed to survive. And to kill."

Although in real life the mild-mannered and sluggish giant squid doesn't chase humans, it has on rare occasions had titanic battles with sperm whales.

* * * * *

The author in the jaws of the giant shark.

No stories about sea monsters would be complete without mention of a shark that makes the great white shark of *Jaws* look like a minnow.

Imagine a creature twice as long as the great white shark, four times as heavy, and a mouth big enough to swallow a cow. Its eight-inch-long teeth could tear apart Tyrannosaurus rex in seconds!

That is the stuff of nightmares. That is 16 tons of terror. And that happens to be the very real Carcharodon megalodon or giant shark. The problem is that scientists believe giant sharks became extinct thousands of years ago. So the question is: Are giant sharks really extinct?

Here's what shark expert, David Stead, wrote in his book, *Sharks and Rays of Australian Seas* in 1963:

> "In 1918, lobster fishermen in Port Stephens refused for several days to go to their regular fishing grounds. The men had been working in that deep water area when an immense shark of almost unbelievable size appeared and lifted huge pot after pot, each containing more than 100 pounds of lobsters. Then it took all the pots and gear. Along with the fishing inspector, I questioned the men very closely, and they all agreed that it was longer than 100 feet. All the men confirmed that the water literally boiled when the monster swam by. These were men used to the sea and all sorts of sharks and whales, and they were convinced it was no whale."

The author concluded, "I have little doubt that in this case humans had actually seen a giant shark, which we know existed in the past in the vast depths of the ocean. While they are probably rare, they must yet be alive."

So whether the giant shark is alive or not has seemingly not yet been absolutely determined.

Now you know more about giant sharks, sea serpents, mermaids, and giant squid than probably anyone else you'll ever meet. You decide what is fact—and what is fiction.

As author Stephen King said, "Monsters are real. They live inside us. And sometimes they win!"

* * * * *

FURTHER READING

Monsters of the Sea, Richard Ellis, Lyons Press, 2001.

The Great New England Sea Serpent, J.P. O'Neill, Down East Books, 1999.

In the Wake of the Sea Serpents, Bernard Heuvelmans, Hill & Wang, 1968.

A Pictorial History of Sea Monsters, James B. Sweeney, Crown Books, 1972.

Mysterious Creatures, Time-Life Books, 1988.

Mystery Monsters of the Deep, Gardner Soule, Franklin Watts, 1981.

P.T. Barnum: The Legend and the Man, A.H. Saxon, Columbia University Press, 1989.

The Search for the Giant Squid, Richard Ellis, Lyons Press, 1994.

Meg (a novel), Steve Alten, Doubleday, 1997.

Sharks in Question, V. Springer & J. Gold, Smithsonian, 1989.

Sharks and Rays of Australian Seas, David Stead, Angus & Robertson, 1963.

CHAPTER 5

SEABIRDS: WINGED WONDERS OF THE OCEAN

Sunrise or sunset, seabirds sail the skies from the Arctic to Antarctica.

"Averno," the Latin word for Hell means "a place without birds." The same could be said for a sea without seabirds. And most of the more than 300 species of seabirds make the oceans and seas their home and food source. As one seabird lover said, "Real birds eat fish."

The great birder, Roger Tory Petersen, told me that "Everyone is born with a bird in their heart. How and when you find it varies with each person." If you haven't yet found that bird in your heart, you might do so by watching the enduring mystique of seabirds on cruises.

Seabirds have to be tough to survive an ocean home. Imagine if you had wings, and your home was a trackless ocean or sea. When it rained, you'd get wet. When it got cold, you'd freeze. When it stormed, you'd get thrown around. And if you didn't catch enough food, you'd starve. No wonder that of the 9,000 bird species on earth, only three percent are seabirds. And only 150 species are able to survive as true ocean wanderers.

But when the going gets tough, the tough get going. One of the most abundant birds on earth is a tiny seabird called a petrel. One flock of storm petrels in Australia was estimated at 150 million birds. And seabirds are the only group of birds to have successfully colonized Antarctica, the planet's most inhospitable continent. So how do seabirds survive—and often thrive—in some of the harshest environments on earth?

To prove that science, in this case ornithology, is largely just organized common sense, what do seabirds need to live on the ocean? Since oceans can be cold and always wet, seabirds need warm, waterproof clothes. Without L.L. Bean, they order from the shop of nature: feathers that they waterproof by covering them with oil from their tail glands. That preening or raking motion also zips up the interlocking structure of their feathers. The interwoven mesh creates a kind of insulated, waterproofed overcoat. No wonder water runs off a duck's back, and geese don't get goose bumps. In freezing, polar areas, seabirds add extra downy feathers and fat layers.

To catch food, seabirds not only have excellent eyes, but an extra set of eyelids that serve as goggles so they can see clearly underwater. A reddish oil in their eyes serves as built-in sunglasses. They also have strong, hooked beaks that hold and tear their prey food.

Seabirds can drink salty ocean water without harm because they eliminate excess salt through glands and membranes in their heads.

They also must be strong fliers if they have to travel long distances for seasonal and scattered food. Many have long wings for near-effortless soaring.

Seabirds are the planet's champion navigators. They can fly thousands of miles day and night with pinpoint accuracy. Nobody knows exactly how they do it. Most experts believe that their brains have a guidance system that serves as a compass using earth's magnetism and the sun and stars. Others believe they use sight and scent cues. It's amazing when you realize that they do it with a brain the size of a pea. In contrast, cruise passengers with their three-pound brains can be on a ship for a week and still get lost. Makes you wonder just who really is birdbrained.

Out of the some 300 species of seabirds, I've focused on those that are most often seen by cruisers in the Caribbean, Alaska, Mediterranean, and South Seas regions.

* * * * *

Soaring gulls are a common ocean sight.

Gulls are the ocean's most familiar beachcombers. To prove that even scientists have a sense of humor, ornithologists gave gulls the family name of Laridae, after a Roman nymph who had her tongue cut out by Jupiter because she talked too much. With or without tongues, gulls are still nonstop talkers!

The 40 species of gulls typically hang out close to shore where they dine on fish, frogs, clams, and insects—live or dead. They also like to follow ships for food. Cruisers say you can tell what ship has the best food by how many gulls follow it.

Gulls are very opportunistic so if they can't find an easy meal at sea, shore, or in fields of corn and grain, they are just as happy hanging out at the local garbage dump. No wonder many nations protect them because they are superb scavengers.

Gulls are also very adaptable. They are one of the few creatures able to drink both fresh and salt water so they can live on lakes or oceans. They also can walk, swim, and fly with equal skill. Some even claim that gulls are so adaptable that during dust storms, they will fly backwards to keep the dust out of their eyes.

So perfectly proportioned are their wings that the Wright brothers studied and measured them as guidance in designing their airplanes. Yet ironically, a Wright brothers plane crashed because it collided with a gull.

Gulls are also full of tricks, and known to fly off with golf balls. In London during World War II, German sabotage was suspected at a munitions factory because of all the broken windows. When guards were posted, they discovered that gulls were having fun dropping nuts and bolts through the skylights.

One scientist was sure that gulls were super-smart because they would fly high and drop oysters and clams on rocks and paved areas to crack them open. Further study revealed that gulls dropped just as many clams on soft mud and sand. He concluded that gulls couldn't tell the difference between soft and hard surfaces.

An old belief said that gulls must be protected because they carry the souls of sailors lost at sea. Yet in parts of the world, they eat gulls, and consider a gull egg omelet as special as caviar.

Instead of eating gulls, Utah built a tall, gilded monument to them in Salt Lake City because they saved the farmers from ruin by eating the grasshoppers that were destroying their crops. As a result, gulls are Utah's state bird.

Male gulls typically fall in love with smaller-sized females, and most mate for life. But some species seem to have a high divorce rate since one-in-four split up after the first year. On one California island, 14-percent of the female gulls performed courtship rituals together, paired off to nest, and even tried to incubate eggs.

Gulls have a strong attachment to their birth place, and tend to return to the same nest sites year after year. They have a rigid, highly developed pecking order. The oldest rules. Gulls know each other individually, and will chase-off strangers.

But gull experts seem to have more questions than answers. For instance: Why do gulls suddenly squabble after long stretches of peaceful togetherness? Why don't they seem to know or care whether they hatch their egg or another? And why do gulls take five days to identify their own offspring? When it comes to gulls, mysteries abound.

* * * * *

Few have heard of American poet Dixon Merritt, but most everyone knows his poem:

A wonderful bird is the pelican,
His bill will hold more than his belican.
He can take in his beak
Food enough for a week,
But I'm damned if I see how the helican.

Merritt wrote a memorable limerick, even if he got his facts wrong. Pelicans don't store food in their beaks or expandable pouches, but if they could it would be big enough to hold 25 pounds of fish or a month's food supply.

17th-century bird experts also got it wrong when they claimed that pelicans pecked their breasts to feed their babies with fresh, warm blood. That belief caused pelicans to become symbols of Christ.

Of the seven species found in the tropical and temperate regions of both hemispheres, only the brown and American white pelicans are found in the United States and Canada.

Brown pelicans are clumsy divers, but expert fish catchers.

Brown pelicans are spectacular fishermen. After spotting fish from 100 feet above water, they fold their wings, stick out their feet, and plummet awkwardly down. As they hit the water, inflatable air sacks under their skin cushion a blow that could sting like a belly flop. The impact sends ripples that stun nearby small fish. With bills open, they scoop up their prey, then bob to the surface, drain their pouch of water, and enjoy a fresh fish supper. Said one observer, "Pelicans fly with the dignity of Roman Senators, but they dive with the grace of a cow."

Unlike brown pelicans, American white pelicans are team fishermen. They form a line or horseshoe formation, then paddle toward shore herding fish into the shallows where they submerge their bills, open wide, and scoop up their meals.

Pelicans often fly in long, slanting or V-shaped formations. Along shorelines, they glide effortlessly by riding thermal updrafts. Like any good pilot, they take off and land into the wind.

During breeding and nesting, pelicans gather close to the water in large colonies to build stick and grass nests in trees or right on the ground. So many pelicans nested on one island in San Francisco Bay that Spanish explorers named it for their word for pelicans—Alcatraz.

In the words of ornithologist, Frank Chapman, "No one can look a pelican squarely in the eye without being impressed by the bird's reserved, grave dignity. The same patriarchal bearing in a man suggests the wisdom of sages and prophets."

But after studying pelicans at length, Dr. Chapman concluded that they looked a lot wiser than they were. He decided they had a limited IQ because life was too easy and not challenging enough since food was plentiful, climate ideal, and they had no real enemies.

That, of course, was before humans shot pelicans for their feathers and unknowingly poisoned them by extensive use of DDT. When DDT was banned in 1972, pelicans made a dramatic comeback. But they still face the dangers of fish hooks, monofilament lines, and human disturbances that result in the abandonment of their eggs.

* * * * *

Cormorants fish around the world.

More than 25 species of duck-like cormorants are found along temperate and tropical coastlines. They have long necks, bodies, and wings along with slender, hooked-at-the-tip bills that are ideal for catching and holding their fish prey. The word cormorant comes from its Latin name "Corvus marinus" which means "sea crow."

Thanks to air sacks they control like submarine ballast tanks, cormorants can bob on the surface like a cork, cruise through the water with only their head showing, or fast dive to catch fish. Their streamlined body is perfect for swimming underwater so they can dive 100 feet using their feet as propellers. Their eyes are not curved like humans but flat giving the bird superb vision above and below water.

Cormorants catch—and count—fish.

Cormorants are so skilled at catching fish that Asian fishermen put them on leashes (fitted with neck collars to keep them from swallowing), and send them diving. Cormorants can even count. Traditionally, fishermen in China give every seventh fish to the cormorant as a reward. If the fisherman forgets, his cormorant won't dive until he gets his fish.

Cormorants benefit humans in another way. They are known as "the world's most valuable bird" because they produce vast amounts of excrement called "guano" on islands off the coast of Peru. Millions of tons of guano have been mined because it makes the world's best organic fertilizer. Guano can be 150-feet deep with the lower layers deposited 2,500 years ago.

The Incas of Peru so valued guano that they put to death anyone disturbing breeding cormorants. Over the years, guano has brought more wealth to Peru than all its gold and silver. Early sailors, however, had another reason to love cormorants. A favorite meal was cormorant hash.

* * * * *

Terns migrate up to 20,000 miles every year.

Terns are the champion migraters of all seabirds. Streamlined and small, the Arctic tern holds the world record for flying 20,000 miles round-trip every year from the Arctic to the Antarctic. It takes three months of flying each way.

Nearly all terns have forked tails and elongated outer tail feathers. They also have shrill, grating calls. Like most seabirds, terns have webbed feet, but they rarely swim because if they settle in the water too long, they become waterlogged.

Terns use a hover-and-dive technique or surface-skimming to catch flying insects and fish. Some flip a caught minnow in the air, and then gulp it down.

Males looking for mates use their fish catches like humans use orchids and champagne to impress and persuade possible mates. They nest in crowded colonies, and typically lay their eggs in shallow depressions right on the ground, except for fairy terns of the Pacific who don't use nests at all. The female lays and then balances a single egg on the bare branch of a tree. The egg is more round than oval so when the female isn't sitting on it and a wind comes up, it supposedly spins safely on its axis and doesn't fall off. At least, that's what the local story tellers say.

Fairy terns use tree branches for nests.

What is true is that the chicks are born with enormous, fully-developed feet letting them hold on to the branch at birth since they have no nest to cradle in. How do they sleep without falling off? Like many birds, they have muscles in their feet that automatically contract when they fall asleep.

* * * * *

Booby birds do high-dive fishing.

First, you'll notice this bird because of its five-foot wingspan and two-foot bill as it flies by. Then you'll see its streamlined body, and think it's the perfect shape for diving. Soon the action begins as the birds spot a school of fish. Wings fold to their sides as they plummet 60 feet or more into the ocean like feathered missiles. Close-up it sounds like hissing hailstones as dozens hit the water at the same time. Few fishermen put on more spectacular shows than the plunge-diving booby birds in tropic and temperate waters.

75

Boobies got their name from the Spanish word "bobo" meaning dunce or fool. Since they typically don't fly from humans and will perch on ships, they were easily caught—and so boobs.

Blue-footed boobies are colorful courters.

The brightly colored feet of blue-footed and red-footed boobies play a starring role in their mating ritual. Instead of long phone conversations and dinner dates, these boobies fall in love by bobbing their heads and jabbing and pointing their bills skyward. Males then strut and stomp in front of the females. Both wiggle their blue or red feet.

If the courtship works, the female lays one or two eggs either in tree nests or on the bare ground. After the male and female have taken turns incubating for 40 days, the chick or chicks are born.

* * * * *

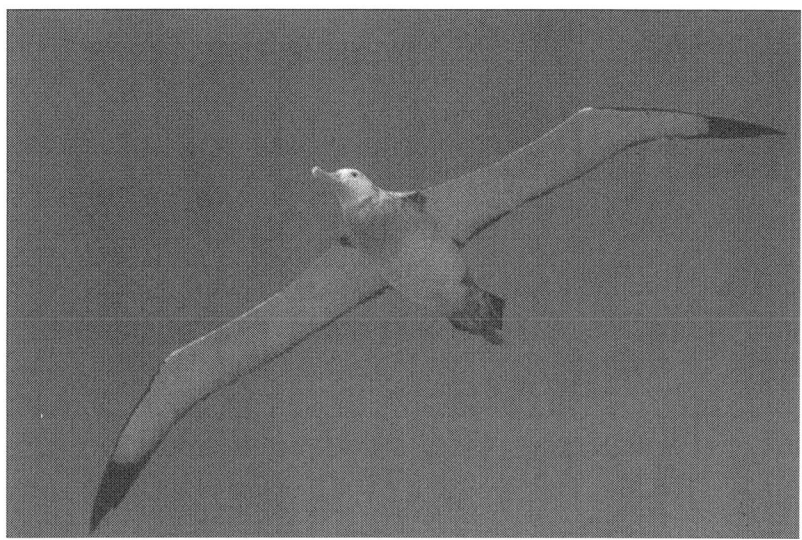

The sky is home for albatrosses except to roost and raise their young on land.

Largest of all seabirds, the 13 species of goose-sized albatrosses typically spend much of their lives high above the oceans. With up to 12-foot wingspans (the longest span of any bird), they are built more to float in the air than fly. They even sleep while soaring, returning to their birthplace only to breed and raise their young. Champion long-distance wanderers, one banded albatross flew around the world in 46 days.

Sailors have long had a special reverence for these aerial nomads of the sea, who follow ships for food when they are not snatching up squid that nightly swim to the surface in huge schools.

Because sailors believed these birds were the reincarnation of seamen washed overboard, albatrosses that followed ships brought good luck while killing them brought bad luck. In Coleridge's famous poem, "The Rime of the Ancient Mariner," the ancient mariner kills one, and his shipmates hang it around his neck as a reminder that it was his fault that they were becalmed in mid-ocean. The incident of shooting the albatross that brought that bad luck actually happened near Cape Horn.

Albatrosses brought plenty of bad luck to U.S. Navy pilots after World War II on the Pacific's Wake Island where thousands nested on the airport's runways. Like the Navy pilots, the albatrosses found the flat runways perfect for running up to takeoff speed before flapping into the air. Collisions were common—and dangerous—as the 20-pound birds smashed windshields, dented wings, and bent propellers.

Navy experts tried everything to chase the birds off. They used flares, smoke, high-frequency sound waves, mortars, bazookas rockets, even stink bombs. Nothing worked until they paved over the nest sites along the runways and leveled the nearby sand dunes that had created uplifting air currents that helped the albatrosses takeoff.

Getting airborne has always been a problem for most albatrosses. They need a long run into the wind to take-off, and plenty of wind to soar once airborne. If calm weather grounds them, they sleep on the water like ducks.

Sometimes albatrosses land on ships where they waddle clumsily around the decks, often unable to takeoff because there isn't enough taxiing space. Sailors nicknamed them "gooney birds" because of their awkward behavior on ships. Paradoxically, these ultimate seabirds can get as seasick as any landlubber when on a rolling ship.

Though world roamers, albatrosses, like most seabirds, breed where they were born—usually less than a thousand feet from where they hatched. Like boobies, they go through elaborate courtship rituals. They bow and bob their heads like country-dance couples, clack their bills, and "moo" like cows. Albatrosses mate for life, and take turns sitting on the one-pound egg. Newborn chicks will gobble down at one sitting four pounds of squid and fish regurgitated by the parents. The fluffy chicks don't fledge and fly away for a year. Some live more than 50 years.

Because 100,000 albatrosses are snagged and drowned every year by the long lines of fishing trawlers, a recent global treaty signed by Australia, Ecuador, South Africa, and Spain now obliges their fishing vessels to reduce seabird snags, protect their breeding grounds, and lower ocean pollution. Hopefully, other nations will do the same.

* * * * *

Frigate birds are riders and rulers of ocean winds.

Frigate birds hang in the wind like kites on invisible strings. Yet they can hover like hummingbirds, swoop like hawks, and glide like albatrosses. They are so adapted for flying that their skeletons weigh less than their feathers. Their eight-foot wingspan is wider than a bald eagle's, yet frigates weigh less than a gull.

There are five species of frigate birds that soar the tropic oceans. They fly as high as 4,000 feet and can travel 1,000 miles without landing. They eat on the wing by snatching small fish from ocean waves, drink on the wing, preen and bathe on the wing by skimming and dipping along calm surface waters, and even doze on the wing.

Appropriately, frigate birds were named for the swift and graceful three-masted sailing ships. Polynesians once trained frigate birds to carry messages between islands. Early seamen called them "man-of-war birds" because of their skill at dive-bombing and stealing fish from other seabirds.

Although true seabirds, frigate birds can't land on water because their tiny feet have no webs, and their feathers are not water-repellent. Nor can they easily move around on land because their legs and feet are too short and small for walking. So when frigates come to coasts or islands to roost, breed, and nest, they perch on cliff ledges, bushes, or trees.

Male frigate birds attract females by inflating their red throat pouches.

Frigates are famous for their dazzling courtship displays. At the start of the nesting season, the black males with white stomach patches settle in bushes or trees and attract females by puffing up their red throat pouches like bright balloons. As the pouch enlarges, the male spreads his four-foot wings and makes loud, gobbling noises to attract a mate.

The white-breasted females fly overhead inspecting and finally choosing their favorite male. Courtship may then include head-rubbing and cooing. Nesting materials are gathered, and the female lays a single white egg. Both take turns sitting on the nest. Two months later, the tiny, naked, blind chick is born and will soon lift its bulbous head to receive its first meal of regurgitated fish.

After the first month, the chick has acquired a fluffy coat of white down. Two months later it fledges and flaps off on a morning wind with the precision of a master pilot. Flight for the frigate bird is clearly what seabirds are born for.

* * * * *

Although they often "fly" over the ocean, flying fish are, of course, not seabirds. But the 48 species are often seen by cruisers in warm waters around the world.

Up to 18-inches long, flying fish don't really fly. They glide for 100 yards or more by using their wing-like fins and vibrating their powerful tails 50 beats a minute to escape larger fish like tuna and mackerel. Sometimes albatrosses and frigate birds will grab them with their hooked beaks just as they skitter above the surface. So tasty are they that flying fish are popular eating in the Caribbean.

And don't be surprised if a wayward flying fish ends up on the deck of your ship. These fish of flight have been found on the decks of ships as high as 30-feet above the ocean surface. The theory is that once airborne, they are swept higher by a strong headwind and plunk down on the ship.

The "hummingbird of the sea" swims more than flies.

* * * * *

So far, we have described seabirds that are seen mostly in tropical and temperate areas around the world. The following group of seabirds are called alcids. Most live along the chilly northern coasts of the Pacific and Atlantic. The 22 species include auks, auklets, murres, guillemots, and puffins. In many ways, these seabirds are the northern hemisphere's version of the penguins of the southern hemisphere.

Their similarities are striking. Like penguins, alcids use their wings to "fly" underwater. Both alcids and penguins use their feet to steer—not propel like other seabirds. And both dive deep for their food. Like penguins, alcids stand erect in crowded colonies. Their feet are typically far back on their bodies. And like penguins, alcids are black and white, have dense, waterproof plumage to keep warm in freezing weather, and many have bright colored bills and head feathers.

But there are differences. Alcids can fly, penguins can't. Interestingly, there was one alcid—the great auk—that was as flightless as penguins. Defenseless against humans, the great auk became extinct a century ago.

So in a fascinating example of evolutionary convergence, alcids and penguins have ended up as look-alikes with similar behavior yet live in opposite but matching habitats on the planet.

* * * * *

The parakeet auklets of the north Pacific are penguin look-alikes.

The most abundant small alcids along the Pacific coast and specially Alaska are the quail-size auklets. They winter on the open ocean, then in spring come to land to dig burrow nests. Both parents share in sitting on the single egg as well as feeding the chick regurgitated small fish and shrimp they have caught by "flying" underwater.

* * * * *

Common murres lay pear-shaped eggs that reduce rolling off their cliff nests.

Murres are typically larger than auklets and summer along the northern coasts of the Pacific and Atlantic. Their chorus of hoarse moans and low murmuring calls that can be heard a mile away likely inspired their name. Because they lay a single egg on precarious cliff ledges, the egg could easily fall off. But a curious adaptation has occurred. Their eggs are markedly pear-shaped so less likely than round eggs to roll over the edge.

Since murres nest by the thousands in jam-packed rookeries, they were easy targets for 19th-century egg collectors.

In just six years, three million eggs were collected and sold in San Francisco for 20-cents a dozen.

* * * * *

Puffins "fly" underwater for their seafood meals.

The most colorful members of the alcid family are the puffins. Some call them "sea parrots" because of their bright, parrot-like beaks. Two species live in Alaska and along the north Pacific coast. The third lives on north Atlantic coasts.

During their spring and summer mating season, puffins dress up as if they were clowns in the circus with red and yellow bills, white chests, and orange feet.

Using their sturdy bills as pickaxes, puffins dig their nest burrows on grassy slopes close by the water. The burrows are filled with an endless racket of harsh, croaking sounds. To keep foxes and other animals from raiding their borrows, puffins often dig twisting tunnels to the nests, and conceal the entrance with a rock. In a few countries, humans kill puffins for feathers and food while restaurants serve up roast puffin.

When the chick emerges, the parents fly off to catch small fish and return with their catch neatly arrayed crosswise on their bills. A chick will eat its own weight daily so there are lots of fish to catch. Alaskan fishermen are hardly pleased when puffins fish by stealing the bait off their hooks.

When the nesting season is over, puffins change their clown colors. By winter, when they go to sea, puffins have dark faces, black bills, and no gaudy feathers. It's as if the party's over, and it's time to get serious about coping with winter on the cold ocean.

* * * * *

Bald eagles catch fish as skillfully as seabirds.

Although not considered a seabird, bald eagles depend on water for their livelihood as much as seabirds. Like seabirds, their main food is fish. They nest in tall trees close to water. Even their scientific name Haliaeetus leucocephalus means "sea eagle with white head."

Found only in North America, bald eagles are more abundant in Alaska than anywhere else with an estimated 15,000 along Alaska's coastal route of cruise ships. On average, there is an eagle nest for every mile of shoreline.

Eagles have symbolized strength and courage since earliest times. They went into battle at the head of Persian soldiers. Their likeness rode on the standards of Roman legions. They served as live rallying points for every regiment in Napoleon's armies. More nations have adopted the eagle as their symbol than any other bird. So it was hardly surprising that the United States chose the bald eagle, a bird unique to North America, as its national emblem.

But it wasn't a unanimous decision. Ben Franklin called the bald eagle a cowardly, carrion eater of bad moral character because it stole fish from other birds. He said it was nothing more than a vulture with a press agent. Franklin's choice for our national bird was the turkey.

The bald eagle, with its seven-foot wingspan, three-foot height, and 15-pound weight, is the largest member of the hawk family—and the largest bird in America after the California condor.

Although Ben Franklin correctly described them as carrion eaters, bald eagles are also excellent fishermen. They can spot fish half-a-mile away. Unlike most seabirds that use a dive-bombing technique, bald eagles glide just above the water, extend their feet, and hook their prey without wetting a feather.

When salmon congregate along streams during summer spawning, you can expect to see lots of bald eagles. Amazingly, they can even swim—using a butterfly stroke with their wings—to bring salmon to shore to eat if they are too heavy to fly off with.

The largest known gathering of bald eagles in the world occurs along Alaska's Chilkat River near Haines. Here in the fall, 4,000 gather in cottonwood trees along the river waiting to feed on the late salmon run.

Bald eagles court in a spectacular way. Their pre-mating ritual is to somersault through the air with their talons locked together. Like other birds of prey, the females are larger than the males, reach maturity slowly, and typically lay two eggs. Often the oldest, strongest chick will peck and kill the younger.

Bald eagles, who mate for life, are prodigious nest builders. They return each spring to the same nest, adding fresh grass and twigs. Some nests can be 10-feet across, 20-feet deep, and weigh two tons—the size and weight of a pick-up truck! Bald eagles aren't fussy about their building materials. Nests have included fishing plugs, light bulbs, even a tablecloth.

Feeding two chicks up to five times a day keep parents busy.

Like all baby birds, the chicks are constantly hungry so the parents are always busy fishing. In three months, the eaglets grow in length from three inches to three feet. By fall, the 10-pound chicks have fledged and left the nest. The black feathered juveniles will not have their distinctive white head and tail feathers until their fourth or fifth year. When the bird was named several centuries ago, the word "bald" meant white.

Although the bald eagle is featured on coins, dollar bills, flags and the Presidential Seal, they have struggled to survive despite past protection. Alaskan fishermen and fox farmers once claimed that bald eagles were killing off their livelihood. So in 1917, the Territorial Government put a bounty on the birds. A hundred thousand were killed before those claims were found to be false in 1953.

Mistaken for golden eagles until their head and tail feathers turn white, bald eagles have been illegally gunned, poisoned, and trapped as killers of livestock.

The heavy fish diet of the bald eagle almost caused its extinction in the lower 48 states. The widespread use of DDT in the 50s and 60s caused build-up in all fish-eating birds. Insects poisoned by DDT were eaten by small fish that were eaten by larger fish—and in turn by eagles. The DDT concentration increased at each stage up the food chain until eagles were laying sterile as well as thin-shelled eggs that cracked during incubation. By the late 60s, only 400 breeding pairs remained in the Lower 48. Fortunately, the banning of DDT and other insecticides in 1972 saved the day. Today, there is a thousand-percent increase in nesting bald eagles compared to 30 years ago.

The lesson is clear. Bald eagles, like seabirds, are more than just birds to be enjoyed. They serve as an alarm that signals trouble in our oceans and rivers. If seabirds are having trouble, the oceans are having trouble. And if the oceans are having trouble, we are having trouble.

* * * * *

FURTHER READING

Seabirds: An Identification Guide, Peter Harrison, Houghton Mifflin Co., 1983.

Birds of the Atlantic Ocean, Ted Stokes, Audubon Society, 1968.

The Life of Birds, David Attenborough, BBC Books, 1998.

Water, Prey, and Game Birds of North America, National Geographic Books, 1965.

Secrets of the Seas, Reader's Digest, 1972.

A Field Guide to the Birds; East of the Rockies, Roger Tory Peterson, Houghton Mifflin Co., 1980.

The Wild Edge; Life and Lore of the Great Atlantic Beaches, Philip Kopper, Times Books, 1979.

Lords of the Air; The Smithsonian Book of Birds, Jake Page & Eugene S. Morton, Smithsonian Books, 1989.

Ocean Birds, Lars Lofgren, Crescent Books, 1984.

The Blue Planet; A Natural History of the Ocean, Byatt, Fothergill, Holmes, BBC Worldwide, Ltd., 2001.

100 Birds and How They Got Their Names, Diana Wells, Algonquin Books of Chapel Hill, N.C., 2002.

Silent Spring, Rachel Carson, Houghton Mifflin, 1962.

CHAPTER 6

SEA OTTERS, SEA LIONS, SEALS

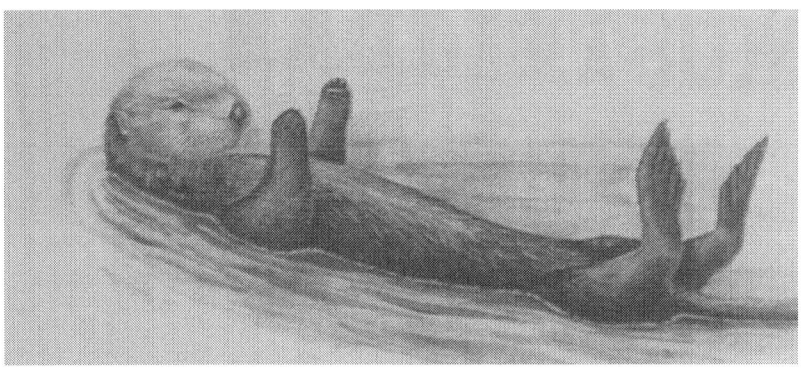

Sea otters wear nature's finest fur coat.

Sea otters (Enhydra lutris), like whales and dolphins, are marine mammals—but with a difference. As marine "newcomers" of only five-million years, they still look a lot like their land relatives. They have legs not flippers, lots of hair, and external ears. But already they have evolved webbed hind feet, a broad, almost paddle-like tail, and ears smaller than land-based otters to reduce drag in water. Unlike land otters, they also eat, sleep, mate, give birth, and nurse at sea.

To keep warm in the freezing waters of Alaska, Japan, and Russia, sea otters have evolved the thickest fur of any mammal with a million hairs per square inch—double that of minks. Warmth is maintained by trapping air in that heavy fur so sea otters must regularly groom their fur. If their hair becomes matted or dirty, it won't trap heat-holding air bubbles, and they can freeze. So hair care is a matter of life-or-death for sea otters.

But if a sea otter's hair is a lifesaver, it has also been a life-killer. When Russian fur traders two centuries ago discovered how thick and lustrous that fur was, they knew they'd struck it rich. Sea otters were naturally trusting and inquisitive, and therefore easily killed by hunters.

One Russian hunter returned after a single trip to Alaska with 10,000 pelts. "They're as tame and innocent as sheep," he reported. Soon the rush was on for what was called "Soft Gold."

The glossy pelts brought phenomenal prices in Asia. Skins bought from Alaskan natives for $3 worth of beads were sold in China for $5,000 each. English and American merchant ships followed the Russians, and by 1911 more than half-a-million sea otters had been killed before protection was given to the remaining 2,000. Populations have recovered, but their survival is still threatened from oil spills and DDT found in Asian shellfish they eat.

Sea otters and fishermen have always had an uneasy alliance because sea otters eat the same shellfish—like abalone—that fishermen depend on for their living. Yet in some parts of the world, fishermen use trained sea otters to catch fish.

Sea otters are Teddy-Bear cute.

Sea otters are called "Aquatic Teddy Bears" because of their cuddly, cute looks and playful behavior. Largest of the weasel family, they are the size of basset hounds and sport bushy mustaches. Like humans, their hair gets grizzly or white as they age.

Lots of female sea otters have skinned noses because the males like to court and make love by biting their noses. Males seem to make love like Mike Tyson boxed!

Females have single pups, and the baby lives on mom's chest for the first month. When in danger, mom will tuck her pup under her foreleg and dive for safety. After six months, pups fend for themselves. But only half the pups will survive because of storms, sharks, bears, and entangling fish nets. Average life expectancy is 10 to 20 years.

Sea otters like to paddle on their backs in dense kelp beds. Groups will often wrap themselves in kelp, put their paws over their eyes, and go to sleep.

When not dozing, they are busy diving for mussels, crabs, clams, and abalone. Cracking open shellfish to get food is a hard job, but sea otters make it look easy. They collect flat rocks, balance them on their chests, and break open the shells by smashing them down on the rocks. One sea otter brought up 54 mussels in 80 minutes, and whacked them against his rock more than a thousand times. When they dive for food, they just tuck their rock under their front leg and swim down. One creative sea otter couldn't find a rock so he used an empty wine bottle. By turning rocks into tools, sea otters are among an elite few animal species that can be called tool-makers.

Because they are relatively small and have no blubber, sea otters need to eat 20 pounds of shellfish a day (30% of their body weight) to sustain the metabolism needed to stay warm. For a human, that's like eating 100 hamburgers a day!

* * * * *

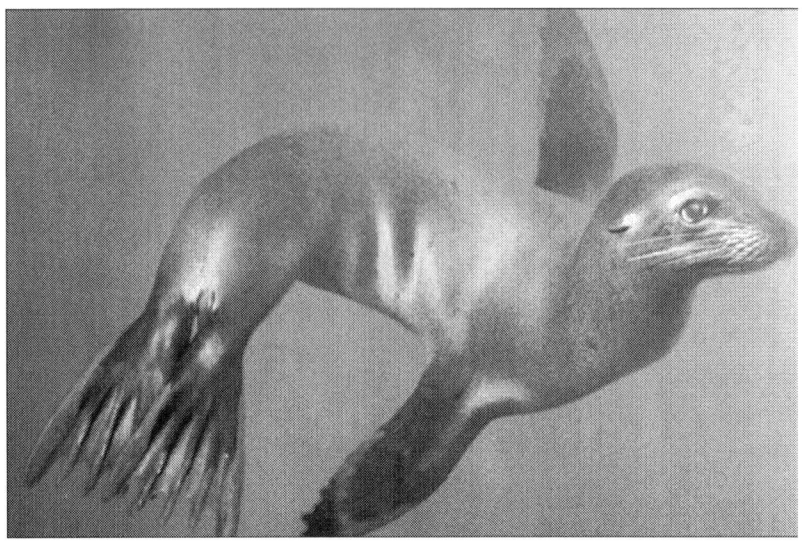

A sea lion takes an underwater swim.

Sea lions (Zalophus californianus) belong to the order Pinnipedia meaning "wing—or fin-footed" because of their paddle-shaped limbs and webbed flippers. Pinnipeds moved to water 30 million years ago, and came from the family that includes dogs and wolves. They still have dog-like heads and like to bark. Their bodies are covered in thick blubber for warmth. Their tiny ears reduce drag. They are still agile on land because their hind flippers go under their body and are used like legs. The trained seals of circuses are actually sea lions.

Like most marine mammals, sea lions have not had happy encounters with humans. They were heavily hunted in the past as their pelts made warm, fashionable coats, their blubber yielded fine oil, and their flesh was tasty.

Fortunately, sea lions are now protected and populations are healthy. They are doing so well in California that they have become pests by lounging on the decks of anchored boats. They eat fish and squid so can be hated by fishermen.

Sea lions use sonar to navigate. They also have excellent eyesight and hearing. So good, in fact, that the U.S. Navy sent some to the Middle East to see if they could defend ships against enemy frogmen and mines.

Sea lions are, however, still linked to land. Rarely do they swim more than a hundred miles from shore. Along coasts and islands from Alaska to Baja California, they form large rookeries for breeding and nursing. Males are highly territorial, and the biggest rule harems up to 50 females.

After a one-year pregnancy comes a single pup that is three-feet long and weighs 14 pounds. Mom's milk is 50% fat, and she nurses for six months.

Sea lion mom and her pup.

* * * * *

Steller sea lions (Eumetopias jubatus) are the largest of the five species of sea lions. At up to 12 feet and 2,000 pounds, they are bigger than grizzly bears. Their range includes coastal areas along the North Pacific—from Japan to the Bering Sea and down the Alaskan coast to California.

Steller sea lions crowd their rookery.

Like other sea lions, they eat squid and eel, and sometimes steal fish from fishermen's nets. They also tangle nets, and sometimes tangle with fishermen. Recently, a Steller thought an Alaskan fisherman was a big, tasty fish so jumped out of the water and pulled the man out of his boat before realizing its mistake and letting him go.

Each year, Steller sea lions return to the islands and shores where they were born. Breeding bulls, distinguished by big manes and thick, muscular necks, arrive at the rookeries in late spring. There is always lots of battling over turf, and females constantly growl and bark loudly.

Pups are born in early summer, and they will suckle until the next pup comes, or longer if mom lets them. I once saw a large female nursing a smaller female that, in turn, was nursing a tiny infant. That's about as close as three generations can get!

A harbor seal and her pup go cruising on an iceberg.

Harbor seals (Phoca vitulina) are the most common of seven species seen along both Atlantic and Pacific coasts. An estimated 70,000 are in Alaska. They can reach six feet and 300 pounds. Their color and pattern vary from dark to light and usually mottled. Many have milky white eyes as they commonly get cataracts.

Harbor seals leave the water only to rest, breed, and give birth. They haul out on rocky islands, sandbars, or icebergs where predators can't easily reach their pups.

Late spring and summer is pupping season. Remarkably, only an hour after birth, pups can swim and dive.

Like all seals, harbor seals live on crabs, shrimp, and fish. And orcas sometimes live on them. As a result, when harbor seals see a pod of orcas, they get out of the water and climb to the closest rock as fast as they can. One seal that was being circled by orcas leaped into a whale-watching boat to escape its predators!

Although Alaskan fishermen used to shoot fish-eating harbor seals for the $3 bounty, current populations seem to be in good shape.

* * * * *

Photographing dozing elephant seals.

Elephant seals (Mirounga angustirostris/Northern & Mirounga leonina/Southern) typically spend their days lying like logs on rocky beaches waiting for people like me to take their picture!

Unlike humans who shed a little of their skin daily, elephant seals shed their skin on land once a year. The process may take a month or more. Their skin peels off in ragged strips that are often strewn up-and-down their breeding beaches. To soothe the itching, they often throw sand over themselves.

Hunted to the brink of extinction by the late 19th-century, the Northern species has rebounded from 100 animals to perhaps 100,000 from San Francisco to Baja California. Northern males can reach 12 feet and 4,000 pounds.

Males of the Southern species, found in the South Atlantic and southern Indian Ocean, can reach 16 feet and 10,000 pounds. Males of both species can be four times heavier than the females, the largest size difference of sexes in any mammal.

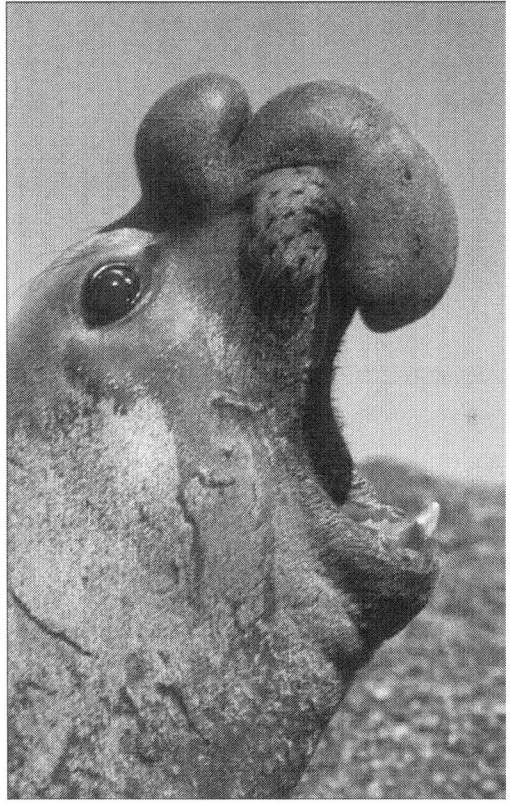

Male elephant seals out-nose females.

As males get older, they develop large, bulbous snouts that overhang their mouths like trunks which gave them their name.

When excited, their trunks become distended and emit bellowing roars.

Bulls practice "tough love" as they fight over territory—and females. They roar and feint, slap each other, and scuffle like sumo wrestlers. They use their trunks as swinging clubs, which can, at times, draw blood.

A breeding male surrounded by hordes of eager females can impregnate up to 50 a season.

But the males pay a heavy price for all that love-making. They are constantly fending off other males while at the same time satisfying the female mobs that come and go constantly. As a result, they never leave the beach so don't eat for two months, and shrink to half their body mass for fear of losing their territory and harem. No wonder males rarely live past their teens, while females breed and live into their twenties.

* * * * *

Since seals have been in water five million more years than sea lions, they have eliminated their external ears in streamlining. Their smaller, front flippers are now used for steering since they get maximum swim-power from extended rear flippers. While humans can swim 4 m.p.h., seals can zoom along four times faster.

Seals are the only mammals that live year-round in the coldest part of the planet—Antarctica—where temperatures can drop to 127 degrees below 0°F. To help them survive, they have a thick layer of insulating fat, and females feed their young the most nutritious milk of any mammal.

Mother seals and their pups use individually distinctive calls to identify and locate each other even in the midst of the acoustic bedlam of crowded breeding beaches. Their echolocation is so sophisticated that blind seals have no trouble catching fish and avoiding predators. But orcas sometimes capture seals and toss them around like a cat plays with a mouse before killing them.

Seals are much clumsier on land than sea lions because they can't put their rear flippers under their body to use as legs. So they awkwardly wiggle like worms along the beach.

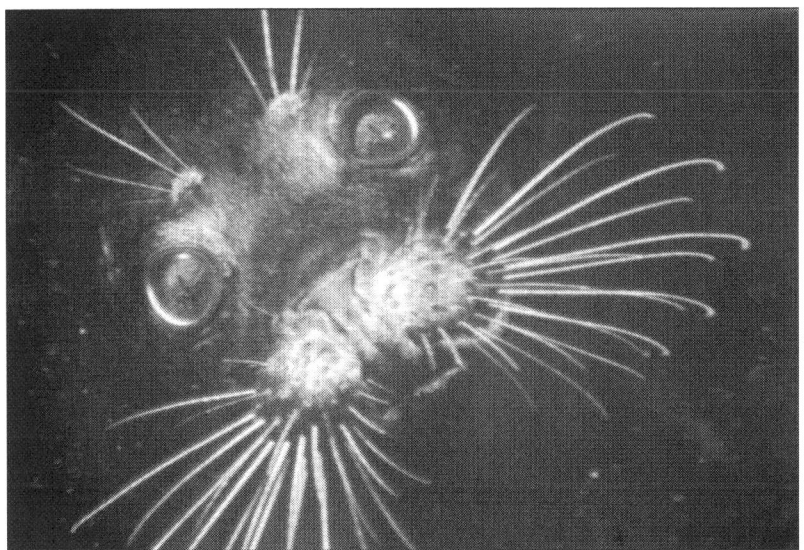

Seals have big, bright eyes.

Some fishermen still consider it unlucky to kill seals because they embody the souls of dead sailors. According to European legend, certain seals who have very large eyes are actually fishermen who were caught in some act displeasing to the gods. So they were made to live in hairy skins forever after and wander at the whim of the winds and tides.

Once in a while, however, such a seal will save the life of a drowning fisherman and so be released from its binding spell. The seal then turns into a beautiful maiden, and the fisherman falls in love and marries her. But there will never be any children. And only the old women of the village know the reason why.

These seal maidens always have dark brown eyes—just like a seal—and soft, beautiful bodies. When the full moon streams through their bedroom window, they lie awake in bed. And you can tell they are different—and not human—because their feet are always colder than an ordinary woman's feet.

* * * * *

FURTHER READING

Sea Otters, Roy Nickerson, Chronicle Books, 1984.

Seals and Sea Lions of the World, Nigel Bonner, Cassell, 1994.

Seals of the World, Judith King, British Museum, 1964.

Seals, Sea Lions, and Walruses, Victor Scheffer, Stanford University Press, 1958.

The Ocean World of J. Cousteau, Vol.10, Danbury Press, 1973.

Audubon Guide to Marine Mammals of the World, A. Knopf, 2002.

Seals and Sea Lions, David Miller, Voyageur Press, 1998.

CHAPTER 7

GREATEST MYSTERIES OF THE SEA

Atlantis destroyed.

According to Einstein, "the most beautiful thing we can experience is the mysterious. It is the source of all true art and science."

And there is no more mysterious place than the ocean. Despite the amazing advances in science, the seas remain our last great unknown. They hold 90% of all life on our planet. Yet we know more about the moon than the depths of the oceans.

Scientists say there are thousands of species in the ocean that have never been seen, much less studied. Since 90% of the ocean deep has never been explored and 70% of our planet is covered with water, that's 140-million-square-miles of unknowns.

140-million-square-miles of mysteries. And the sea is a place where ships, planes, and people disappear!

* * * * *

Although the first and perhaps greatest mystery of the sea took place thousands of years ago, it was really only since 1882 that an American popularized the story to the world.

On the dark night of his 49th birthday, Ignatius Loyola Donnelly sat alone writing at his library desk. He was short and stout. His pants were torn. His coat was threadbare and missing buttons. The cold winds of a Minnesota autumn rattled the windows. Hunched over a flickering oil lamp, Ignatius Donnelly wrote in his diary:

November 3, 1880. Today is my birthday, and a sad day it is. All my hopes are gone. My future is dark and gloomy. My life has been a failure. A mistake. My dreams have so often come to nothing that I cease to hope.

Hard to believe that Ignatius Donnelly was so down on himself. He was an author, lawyer, and at 28 elected Lieutenant Governor of Minnesota. Then as a United States Congressman, he'd been hailed as "the most learned man ever to sit in the House of Representatives."

But now his dream of building a utopian community had collapsed. His farm had failed. He was on the verge of bankruptcy. His despair was as bleak and black as that dark night of his 49th birthday.

Then his eyes fell on a forgotten box of old research papers. A new plan and fresh hope fired his mind. He would organize those haphazard papers into a book unprecedented and controversial.

Five months later he took his manuscript to America's biggest publisher, Harper & Row in New York City. They loved it! Donnelly celebrated by treating himself to a lavish steak dinner—for 35-cents in 1881!

His book was an instant best-seller in America. Then a best-seller around the world. Soon Donnelly was wearing tailored clothes and gold buttons on his new coats.

The title of his book was *Atlantis: The Antediluvian World.* Onetime down-and-outer, Ignatius Donnelly, had launched the modern world's fascination with the mystery of the lost island of Atlantis. His book is still in print around the world.

According to Donnelly's book, the citizens of Atlantis enjoyed a brilliant civilization, and passed on their inventions and achievements to other people as far away as Africa, Asia, and the Americas.

To Donnelly, that explained why the same pillared temples of Atlantis stood in Greece. Why the same pyramids of Atlantis rose in Egypt and Mexico. Why the same seedless bananas and other domesticated plants of Atlantis grew in Asia, Africa, and America. And why their same hieroglyphic writings were used by the pharaohs of Egypt and Mayan kings of Central America.

Donnelly concluded that all of the world's greatest accomplishments in art, architecture, and science originated in Atlantis 12,000 years ago. He claimed that every line of race and thought, of blood and belief, originated in Atlantis. In short, if it exists, it began in Atlantis!

Ignatius Loyola Donnelly had ignited a bonfire of interest in the mystery of Atlantis. Or actually re-ignited interest. Because the story of Atlantis had first been told 2,000 years before by the Greek philosopher, Plato.

Plato had described in detail a dazzling civilization that thrived for centuries. And then one awful day and night it disappeared beneath the sea because of a monster mix of earthquakes and floods. "Listen to my story," wrote Plato, "because extraordinary as it is, it is absolutely true."

According to Plato, Egyptian priests once held documents verifying the history of this lost island that had flourished in the Atlantic Ocean outside the Pillars of Hercules, now called the Straits of Gibraltar. As the capital of a prosperous and powerful nation, its empire stretched from Europe to Asia and Africa.

The island was 300-miles long and 200-miles wide. It had precious metals, fertile fields, and freshwater springs. Farmers tended sweet-scented orchards rimmed by forests where monkeys and elephants roamed. Plato said it was, "the nearest thing to paradise on earth."

On one end of Atlantis rose the capital city, surrounded by a wall of metal. Homes were built of white, black, and red stones. Alternating rings of land and canals circled the city. Wide avenues spanned the canals into the city center. Here rose splendid palaces and temples. The sprawling royal palace had walls of brass and copper that sparkled like fire. Beneath the palace was a dock with underground canals leading to the sea.

The spiritual center of Atlantis was the Temple of Poseidon. Its high walls were plated in gold and silver. Here the king and his nine brothers, who ruled the outlying provinces, made and carried out the laws. Wild bulls were caught for sacrifices.

Several hundred thousand Atlanteans enjoyed gymnasiums, gardens, and public baths with hot water and flush toilets.

Although the people were originally virtuous and wise, successive generations yielded to greed and power. Finally, they declared war on Greece and Egypt to further expand their empire. But after the war, Atlantis was hit by a huge earthquake that triggered a tidal wave that sank the island. Though Atlantis was destroyed, Donnelly believed that some survivors escaped on sailboats, and spread their knowledge around the world.

Since Donnelly's blockbuster book, the fact or fiction of Atlantis has been argued in 5,000 books and articles by oceanographers, historians, archeologists, geographers, even psychics. Dozens of movies have featured Atlantis, and many expeditions have tried to locate the site. After all, Troy was considered pure fantasy until discovered and proved to be very real in the 1870's.

British Prime Minister Gladstone was so impressed with Donnelly's book that he wanted to send a naval expedition to look for Atlantis. Hitler was so convinced that supermen lived there that he dispatched several expeditions hoping to tap into its superpowers.

Much but not all of the searching has focused on the Atlantic Ocean since Plato said Atlantis was outside the Pillars of Hercules. So archeologists studied the ancient, black-stone pyramids on the Canary Islands because they have step terraces with flat tops similar to Egypt's oldest pyramids.

A Swedish scientist recently stated that Atlantis was Ireland because the size and landscape resemble Atlantis. The fact that Ireland never sank was explained as Plato confusing Ireland with a part of nearby England that did sink thousands of years ago.

On the other side of the Atlantic, divers off Bimini Island in the Bahamas spotted in 1968 what looked like a road paved with rectangular blocks of stones. Historian and underwater explorer, Dr. David Zink, concluded after many dives that what was called the "Bimini Road" was the man-made remains of Atlantis. And along that underwater highway were rock foundations for homes, a temple, even a seaport.

But the U.S. Geological Survey said the structures had formed naturally, and were similar to beach rock formations along other shores in the Bahamas as well as Japan and Australia.

Some Atlantis scholars believe that the original Egyptian documents that referred to the Pillars of Hercules really meant a locale in the eastern Mediterranean not the Atlantic. It so happens that one island there, the Greek island of Santorini, was destroyed by the world's biggest volcanic eruption that caused huge tsunamis that flooded nearby islands and rattled buildings as far away as Egypt. Today, only the broken rim of the island remains. Like Atlantis, Santorini had a thriving, sophisticated civilization. Even the buildings were made of the same black, white, and red rocks as Atlantis. But Santorini is a fraction of the size of Atlantis and never was a power center.

Many say a better candidate is Crete, largest of the Greek islands and center of the ancient and sophisticated Minoan empire. Like Atlantis, Minoans built magnificent buildings and boasted hot and cold running water and flush toilets. Like Atlantis, a popular sport was catching wild bulls.

The problem is Crete never sank in the sea, but its Minoan civilization was destroyed by 200-foot high tsunamis and buried by 20 feet of volcanic ash from the Santorini eruption. Another problem: the Santorini explosion happened ten thousand years after Plato's 12,000-year-old Atlantis sinking.

Problems, however, have never discouraged Atlantis searchers. Some believe that what Egyptians called the "Real Sea" meant any large body of water. So Atlantis could be just about anywhere! Over the years, it has been "located" off Spain, Germany, Scandinavia, Iceland, Turkey, Russia, Nigeria, Mexico, Japan, and South America.

A few have even suggested Arctic and Antarctic locales. In 1982, a colleague of Albert Einstein argued for a South Pole Atlantis. Professor Charles Hapgood claimed satellite imagery revealed volcanoes under the ice that ringed a large plain. The professor further believed that unique magnetic emissions could be caused by massive walls made of metal, just like those that circled Atlantis. The problem is that geologists say the South Pole has been covered by ice for a million years.

Of the many that have searched for Atlantis, none was more determined than British explorer, Colonel Percy Harrison Fawcett. As a British Army surveyor, Fawcett spent years mapping the jungles of South America. He'd battled piranhas, poisonous spiders and snakes, vampire bats, and crocodiles.

Colonel Percy Harrison Fawcett.

During those years, Fawcett was inspired by early Portuguese explorers who had described a lost city in the jungle of Brazil. Like Atlantis, it was built on a large plain surrounded by mountains. Like Atlantis, it had been destroyed by an earthquake. There were huge blocks of stone carved with letters that once formed arched entrances plus a temple, canals, fountains, circular roads, and walls of colored frescoes.

So when Fawcett retired in 1925, he and two others trekked into the jungle to find Atlantis. His biggest financial supporter was John D. Rockefeller, Jr. After writing to his wife about rumors of an ancient city on a lake, Fawcett and his companions disappeared. Their remains were never found. Then in 1936, an Irish psychic, Geraldine Cummins, claimed she had received mental messages from Fawcett saying he'd found Atlantis. He told her it was so idyllic he refused to leave. Twelve years later, the psychic reported that Fawcett had told her he was dead.

To sum up, most historians believe Atlantis never existed as Plato described it. He had written a morality tale about the decline and fall of a powerful but corrupt empire. Plato made it up to remind Athenians—and future superpowers—of the dangers of arrogance, greed, and power.

But whatever you believe about Atlantis, three facts are indisputable. One: According to a recent poll, nearly half of all Americans believe Atlantis—or a similar ancient, advanced civilization—once existed. Two: If Atlantis is ever found, it would be the greatest discovery of all time. Three: A very real Atlantis already exists on a very real paradise island. So happens it's a $125 million resort called Atlantis on Paradise Island in the Bahamas. It has a thousand luxury suites, a dozen gourmet restaurants, gambling, and pools with underwater ruins. The original Atlanteans would surely feel at home there!

* * * * *

Far more frequent than cities and islands swallowed by the sea are the thousands of ships and sailors that mysteriously vanish without a trace. Some reappear. Some don't.

Take the *Mary Celeste*, for instance. It was a cold, gray November morning in 1872 when this 100-foot brigantine sailed down New York City's East River headed for Genoa, Italy.

There was no reason for veteran Captain Benjamin Spooner Briggs to think the voyage would be anything but routine. The ship had an experienced crew of seven plus Captain Briggs' wife and two-year-old daughter. Mrs. Briggs frequently sailed with her husband.

Captain Benjamin Spooner Briggs.

Captain Briggs was the son of a sea captain, three of his four brothers had also gone to sea, and he'd rounded Cape Horn as a teenage deckhand. Friends said if you cut Briggs, salt water would flow from his veins.

The Captain's only worry was that his ship was loaded with 1,700 barrels of industrial alcohol. He'd never before carried flammable cargo. So he double-checked the barrels for leaks since fumes could catch fire and explode.

But if Captain Briggs had been superstitious, he would have worried a lot. His ship had a long history of bad luck. When she was launched 11 years before as the *Amazon*, her first captain had died on the maiden voyage. Later, the ship crashed into and sank another vessel. Then she ran aground in a storm after numerous other accidents. Finally, the ship ended up in New York where new owners made extensive repairs and re-named her the *Mary Celeste*. This was her first voyage. Sailors in those days believed changing a ship's name was sure to bring even more bad luck. Captain Briggs just laughed at the silly superstition.

As the wind picked up, Captain Briggs steered east for the Azores, his first landfall across 2,000 miles of the Atlantic. Halfway across, Briggs wrote in his log, "2 more barrels of alcohol split. Hull leaking some, but pumps adequate. Weather still rough."

Two days later he noted, "November 25. Azores 6 miles distant." There was no mention of bad weather, illness, or any ship problems. It was the Captain's last entry.

Two weeks later, a passing freighter spotted the drifting *Mary Celeste*. The main sails were furled. The smaller sails were tattered and flapping. No one was on deck. Signals went unanswered.

When the crew of the freighter boarded the *Mary Celeste*, the only sound was the slap of water against the hull. "Ahoy," they shouted. No answer. There wasn't a sign of life or struggle. No real damage. No ramshackled cargo. And not a trace of the Captain, his wife, daughter, or crew. Sheet music still stood on the piano. Below deck was a six-month supply of food and water, but everything was wet. In the hold, four feet of water sloshed against the barrels of alcohol. The sailors had even left behind their pipes and tobacco. No seaman would forget his pipe unless in a desperate hurry. The only thing missing was the lifeboat and some navigational equipment.

The ghost ship, Mary Celeste.

After pumping the hold dry, the men sailed the *Mary Celeste* 600 miles to Gibraltar to claim salvage rights. The ship arrived on an appropriate December day—Friday, the 13th.

The crew, passengers, and lifeboat would never be found. The *Mary Celeste* would return to service. But after many deaths, accidents, and owners, she finally ran aground and self-destructed in Haiti in 1885—exactly 13 years after being found adrift and abandoned. Appropriately, Haiti is called "the Island of Voodoo." And in less than a year, her last captain was dead, one of his sailors went mad, and another committed suicide. The curse on the *Mary Celeste* had continued non-stop to the very end—and beyond!

So what happened to the crew and passengers of the world's most notorious ghost ship? Since there are no witnesses or records, it's anyone's guess. An exhaustive Naval Court of Inquiry in Gibraltar found no answers. It was the first time in its history that the Court could not come to any conclusion.

But that hasn't stopped so-called experts from filling hundreds of articles, books, and movies with their theories. Blame was put on everything from mutiny and murder to sea monsters, seaquakes, and giant squid.

The unsolved mystery so fascinated a young, struggling doctor, who wanted to become an author, that he wrote his first, published story based on the phantom ship. The story became an instant best-seller, and the author went on to fame and fortune. His name was Arthur Conan Doyle, creator of Sherlock Holmes.

Certain facts, however, seem indisputable. Captain Briggs had a reputation as a skilled seaman, extremely honest, and a teetotaler. He was revered as a courageous Captain who would never desert his ship except to save lives.

The most reasonable explanation seems to be that the Captain and crew took to the lifeboat thinking the ship was sinking or its cargo of leaking alcohol was about to catch fire or explode. We do know that when the cargo was unloaded, nine barrels of alcohol were empty. Once in the lifeboat, a storm capsized and sank the lifeboat and everyone drowned. Records in the Azores indicate there had been a storm at that time with fierce winds and torrential rain.

That may be the best explanation for what might have happened. Problem is no matter how reasonable, it is still just a good guess without witnesses or evidence. So the *Mary Celeste* remains one of the greatest mysteries of the sea.

If the *Mary Celeste* is one of the greatest unsolved mysteries of the past, another great mystery is still going on!

* * * * *

At 2 p.m. on December 5, 1945, five Navy torpedo bombers roared off the runway at Fort Lauderdale's Naval Air Station. The 14 airmen of Flight 19 were on a routine training exercise. They were to fly 60 miles over the Atlantic to the Bahamas, do some low-level bombing, then change course three times, and return to base.

The first leg of the mission was uneventful. But then flight leader, Lieutenant Charles Taylor, a veteran combat pilot, radioed, "This is an emergency...I don't know which way is west." A frantic Taylor reported that both his compasses had stopped working. "Everything is wrong," he said, "even the ocean looks strange."

The day had started sunny and clear. But now a cold front brought in thunderstorms, 40-mile-an-hour winds, and 30-foot waves. Then seeing islands below, Taylor thought he was over the Florida Keys. So he led his squadron northeast toward what he thought was the Florida mainland. In fact, the islands were not the Florida Keys, but the Bahama Keys. He was leading the planes further out to sea.

After an hour, Taylor realized his mistake and radioed Fort Lauderdale, "We'll fly west until we hit the beach or run out of gas." Later, not seeing land, Taylor radioed again, "What course are we on now?" Then silence...forever.

A massive sea-and-air rescue was launched, but the mystery only deepened when a Navy search plane with its crew of 13 also disappeared. For five days and five nights, hundreds of planes and ships scoured a quarter-of-a-million square miles of ocean. No remains were ever found. No wreckage. No oil slicks. No life vests or rafts. No nothing! Six planes and 27 men had vanished as completely as if they'd flown to Mars. A Naval Board of Inquiry concluded, "Causes unknown."

So began—but did not end—the greatest of all modern mysteries—the Bermuda Triangle, a notorious half-a-million square mile patch of the North Atlantic, framed by Bermuda, Miami, and Puerto Rico.

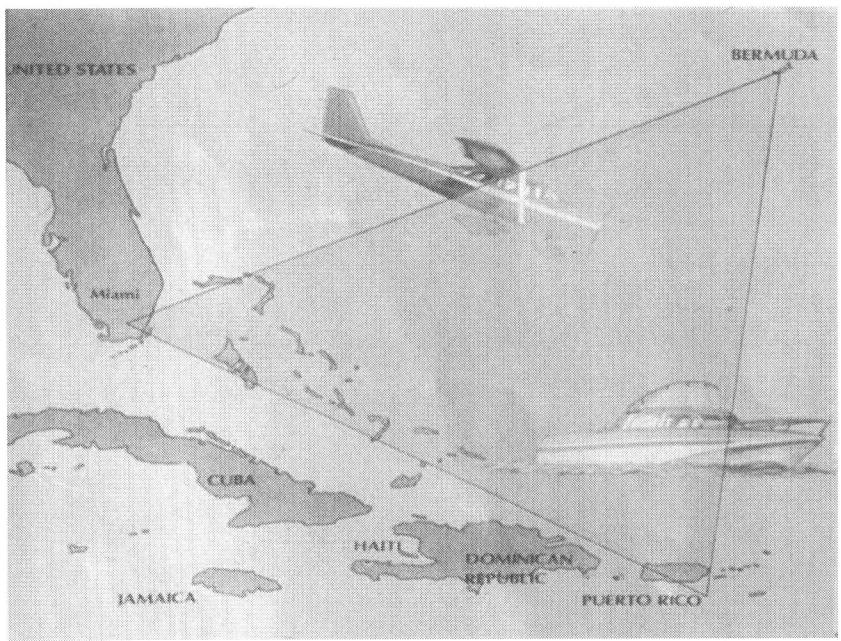

The Bermuda Triangle.

Three years after Flight 19 vanished, so did a four-engine airliner with 31 aboard. After finding no remains, the official report concluded, "What happened will never be known...it remains an unsolved mystery."

Months later, two more airliners with 50 on board vanished. No distress signals. No bad weather. Despite massive, two-week searches, no survivors or wreckage were ever found.

A year later, six more planes vanished. Once again, no survivors or wreckage were found.

And on...and on. An average of five planes a year. Everything from a huge C-119 Flying Boxcar to tiny Piper Cubs went missing in the Bermuda Triangle.

To deepen the mystery, pilots reported strange happenings: Their instruments went out...their compasses spun wildly...their radios sputtered off and on. Some said they'd been followed by unidentified objects. Several had been surrounded by thick, yellow fog and tunnel-shaped clouds that rotated and glowed white.

Not just planes disappear in the Bermuda Triangle. Going back centuries, hundreds of ships and their crews have vanished without a trace.

When Columbus sailed through the area, a bolt of fire smashed into the sea, his compasses malfunctioned, and a strange light lit up the sea one night.

When two U.S. warships with crews of 300 disappeared in the late 18th-century, terrified sailors started calling the area "The Devil's Triangle" and "Sea of Doom."

Captain Joshua Slocum was the first person to sail solo around the world in 1898. During the 40,000-mile voyage, the super-experienced skipper had battled pirates, Cape Horn, and countless, life-threatening storms. He'd faced every danger a sailor could face—and always won. Yet after conquering the world, he set off on his annual winter cruise to Jamaica. He sailed into the Bermuda Triangle, and was never seen again.

In 1918, one of the largest vessels afloat, a 540-foot-long Navy supply ship with 300 sailors vanished north of Puerto Rico. No S.O.S. calls. Despite an intensive search, no survivors or the ship were ever found. And the *U.S.S. Cyclops* was almost as long as two football fields!

Experts say that when a ship that size goes down, objects usually float free. Oil leaks bubble to the surface for as long as a year or more. But no wreckage or oil ever surfaced.

Investigators concluded that, "The disappearance of the *Cyclops* is one of the most baffling mysteries in the annals of the Navy. All attempts to locate her have failed."

Compounding the mystery, two of her sister ships disappeared on almost the identical route. The massive ships and their crew just evaporated.

Even cruise ships have met the mystery. On April 3, 1974, the *QE-2* was cruising through the Bermuda Triangle when suddenly she lost all electrical power and lay becalmed. At the same time, she also disappeared from the radar of a Coast Guard cutter following her. Later, *QE-2* officials blamed the problem on an oil leak.

Recently while my cruise ship was going through the Bermuda Triangle, the ship's doctor, an experienced scuba diver, almost died because his life vest mysteriously inflated forcing him to surface too fast. Hours later, a passenger fell off his balcony in the middle of the night setting off a frantic search by the ship's tenders and a Coast Guard cutter and helicopter. Luckily, the man was found the next morning—cold but alive.

So what causes this mysterious death trap? Is the Bermuda Triangle a supernatural force that swallows hundreds of planes, ships, and people?

Theorists have blamed everything from giant rogue waves to super-powerful whirlpools and electro-magnetic wind storms. Even kidnapping by extra-terrestrial beings! Others are convinced the villains are giant gas bubbles that burst to the surface—and are powerful enough to sink ships. The bubbles come from pools of methane gas under the ocean floor that escape because of earthquakes or other seismic tremors.

Not surprisingly, most scientists say the reasons are natural—not supernatural. After all, the half-million-square-mile area is notorious for furious and fast-changing weather.

Further, the Bermuda Triangle doesn't have any more missing planes and ships than any other busy ocean region. That opinion is backed by records kept by the U.S. Coast Guard and Lloyd's of London.

Even though experts knew about where the 882-foot *Titanic* sunk in 1912, the world's largest moving object at the time was not found for 73 years at a depth of 2.5 miles.

An added factor: Because of the high number of recreational pilots and skippers in the Florida-Bahamas area, the chances of human error are high. Some boaters can't read navigation charts. Some don't even have them! Not too long ago, the Coast Guard got a call from a lost boater. "Where is the island of Bimini?" he asked. "I can't find it on my chart." "What chart are you using?" asked the Coast Guard. Replied the boater, "I have my dictionary open to the world atlas."

Despite all the experts and explanations, the mystery of the Bermuda Triangle is as strong today as it was centuries ago. At least for some!

* * * * *

Finally, a mysterious phenomenon that amazed—and terrified—sailors from the beginning of time.

Every so often when the sun sets at sea, it turns a brilliant emerald green. Because it was so rarely seen, those who never saw it said it was pure fantasy.

When Jules Verne, author of *20,000 Leagues Under the Sea*, wrote a book called *The Green Flash* in 1882, sightings and interest exploded.

Experts quickly came up with many explanations. One said the Green Flash was caused by the light of the setting sun passing through crests of waves that allowed only green light to shine through. The problem was others had seen the Green Flash on land, even waterless deserts.

Another expert said the green color was produced in our eyes because the after-image of a red setting sun is green. But that explanation proved wrong when the Green Flash was photographed on color film.

The Green Flash.

The correct explanation came decades later when scientists realized that the changing colors of a setting—or rising—sun are like overlapping disks of different colors at slightly different heights. The red sun or disk at the bottom disappears first, followed by green, blue, and violet. But the blue and violet colors are blocked at the horizon by the atmosphere. So the last color seen is the top of the green disk.

Despite its name, there is no flash of light. The flash only means it happens as quick as a flash, usually a second or less. However, in Norway, the Green Flash lasted 14 minutes. And Admiral Richard Byrd on his South Pole Expedition in 1929 saw the longest Green Flash on record—35 minutes.

So where is the best place to see the Green Flash? Any sea level place with a low, straight horizon and a near cloudless sky. In other words, right on a cruise ship at sunset.

There are extra good reasons to look. Islanders believe that seeing the Green Flash guarantees you true love. Others say that after seeing it, you can read the thoughts of others. Cynics say lots of rum punches increase your odds!

So there you have just a few of the most famous mysteries of the sea: The amazing phenomenon of the sun turning green that once terrified and now just amazes viewers; a lost city that ruled much of the world before vanishing beneath the water; the puzzle of a ghost ship with a missing crew never found; and a triangle of ocean that swallows planes, ships, and sailors with either natural or supernatural powers.

Are those stories fact or fiction? I don't know. Perhaps that's why they have become the greatest mysteries of the sea.

* * * * *

FURTHER READING

The World's Greatest Unsolved Mysteries, Damon Wilson, Barnes & Noble, 2006.

Mysteries and Monsters of the Sea, Frank Spaeth, Gramercy Books, 2001.

Mysteries of the Ancient World, National Geographic, 1979.

The 70 Greatest Mysteries of the Ancient World, Brian Fagin, Thames & Hudson, 2001.

Imagining Atlantis, Richard Ellis, Vantage, 1998.

Atlantis: The Antediluvian World, Ignatius Donnelly, Harper & Row, 1971.

Atlantis: The Truth Behind the Legend, A.G. Galanopoulos, Bobbs-Merrill, 1969.

Ghost Ship; The Mysterious Story of the Mary Celeste, Brian Hicks, Ballantine, 2004.

Without a Trace, John Harris, Atheneum, 1981.

The Bermuda Triangle Mystery Solved, Larry Kusche, Harper & Row, 1975.

Into the Bermuda Triangle, Gian Quasar, McGraw Hill, 2004.

The Bermuda Triangle, Norma Gaffron, Greenhaven Press, 1988.

The Bermuda Triangle, Aaron Rudolph, Edge Books, 2006.

CHAPTER 8

PIRATES: THE ORIGINAL TERRORISTS OF THE SEA

A typical pistol-packing pirate.

Pirates are surely the wildest, most colorful creatures of the ocean even if they don't live in water. One historian wrote that a pirate had the devil for a father and was suckled by a pig for a mother.

Whatever their origin, many words we use today originated with pirates. "Letting the cat out of the bag" began by pirates and seamen describing the removal of the cat-of-nine-tails whip from the bag to flog a victim. "In a pickle" goes back to when pirates had their backs rubbed with salt and vinegar after a flogging. The expression "to bite the bullet" came about because when surgery was performed—without anesthetic—pirates were given bullets to bite down on during the pain. The word "hooker" originally was used by pirates to describe a poor sailing ship that hooked to the right or left. A phrase I heard often as a youngster, "The bogeyman will get you if you don't watch out," originally referred to the dreaded Bugi pirates of Indonesia. And thank pirates every time you barbecue because it was popularized by Caribbean pirates who were called buccaneers from the French word "boucaner" meaning "to smoke meat."

Because pirates in the town of Tarifa at Gibraltar demanded payments from all ships passing through the Straits, Tarifa gave us the word for taxation—tariff. And thank a pirate for Thanksgiving because the Governor of Massachusetts was so happy capturing Captain Kidd that he made Thursday, November 23, 1699, the first official Thanksgiving holiday. So pirates—not pilgrims—gave us Thanksgiving.

We know—or think we know—all about pirates thanks to Hollywood movies, romance novels, and Treasure Island. Like so many things, the facts about pirates are not only stranger, but more fascinating than the fiction that surrounds them.

It's been said that piracy is the third oldest profession after prostitution and medicine. Ever since humans have had ships, there have been pirates. After all, a pirate is really just a burglar on a boat. So whenever and wherever the rewards of crime are worth the risks of punishment, there will be pirates.

There are 6,000-year-old clay tablets in Iraq of all places that detail deeds of pirates. Our earliest laws, written on stone 2,000 years ago, included laws against piracy. So pirates really are the original terrorists of the sea.

The name "Viking" originally meant "pirate" so there were Norwegian and Danish pirates as well as Arab and Chinese pirates. The most successful medieval pirates gilded their ships with gold and plated their oars with silver.

During Roman Empire days, pirates captured and held a young Julius Caesar until a large ransom was paid. St. Patrick was also kidnapped by pirates in England, taken to Ireland, and sold as a slave.

Ancient Italian pirates even stole Santa Claus—actually the bones of St. Nicholas, the Turkish bishop who inspired the Santa Claus story.

Abraham Lincoln, while piloting a flat boat down the Mississippi, was attacked by river pirates. He escaped, but ever after carried a pirate-cut scar on his ear.

In 1579, Englishman Sir Francis Drake ambushed just one Spanish galleon in the Pacific and grabbed 1,300 bars of silver, 14 chests of gold, and bags of jewels including an emerald as big as a baseball. It was all worth 25-million dollars—more money than his Queen, Elizabeth the First, spent to run her entire government for a year!

The heyday for pirates was the 17th and 18th-centuries when thousands flocked to the Caribbean like ants to a picnic. The weather was delightful, and the many islands provided hideaway harbors, fresh water, and food. The Governor of Jamaica complained to the King of England that so many men were pirates that it was impossible to find crews for merchant ships anywhere in the West Indies.

One pirate in Barbados captured his booty without even using a ship. Sam Lord just strung lanterns high in his palm trees at night so ships thought it was the harbor and crashed on his reef. Then he rowed out for the loot. Pirate Lord did so well he built a castle and stuffed it with a million dollars worth of stolen art and antiques. When his castle was a hotel, you could reserve Lord's bedroom and sleep in the pirate's four-poster bed.

Although gold, silver, and jewels were the most coveted, pirates took anything of value from soap and sails to rigging and rum. They often came ashore to sack ports and grab booty before it even got loaded on ships.

Sometimes entire villages were designed to foil pirates. If you've been to the Greek island of Mykonos or Lisbon's Alfama district and been lost walking their streets, you can blame pirates because the narrow, twisting lanes were designed to confuse invading pirates.

From time to time, pirates found it profitable to offer their services to nations at war. They were called "privateers" and formed a kind of private navy which was a bargain-basement way for nations to fight enemies.

Who were these people we call pirates? We know they had to be elusive. They came out of the blue, attacked, looted, and then vanished. They left no memorials or personal belongings behind. And they weren't much for record-keeping or writing home.

If you believe the books and movies, pirates were handsome, brave, daring, and romantic. They sailed around in rakish black schooners flying the Jolly Roger. They wore black eye patches, were armed to the teeth with pistols and cutlasses, and made their enemies walk the plank. They buried their loot on islands beneath an X marked on a secret map and sang "Yo Ho Ho and a Bottle of Rum" while the parrot on their shoulder screeched "Pieces of Eight."

Far from glamorous and romantic, the life of a real pirate was hard and cruel, as it was for all seamen in those days. Most were disgruntled, out-of-work sailors. Others could be deserters from Navy and merchant ships, murderers and thieves on the run, and runaway slaves. There were even a few black pirate captains. All hoped to get rich quickly. While an English clerk in the 18th-century made $1,000 a year, pirates might make a half-a- million on a single voyage.

A few pirates were young men from good families who thirsted for adventure. As one pirate wrote, "In honest work there is low wages, hard labor, and poor food. A pirate's life is plenty and pleasure, ease, liberty, and power. So my motto shall be, A short life, but a merry one." That motto was mostly true. Most pirates scarcely survived a decade. Some careers were over in two years, and very few retired healthy and wealthy.

Most pirates were illiterate seamen in their twenties or younger. An 18th-century wanted poster in Virginia described some typical pirates: "Tee Wetherly, short, very small, blind in one eye, about 18... John Lloyd, rawboned, very pale, remarkable deformed in the eyelid... Thomas Simpson, small and much squint-eyed, 14 years old."

The renowned pirate Henry Morgan was described by an eye-witness as "bow-legged and sallow with yellowish eyes and jutting belly who sat up late at night drinking too much."

In fact, pirates didn't shave or shower, and they certainly didn't use floss! Most had bad breath, bad teeth, and feet that were rotting and stunk. So much for Errol Flynn and Johnny Depp look-alikes!

Favorite pirate ships were the large square-riggers with 40 cannons and room for crews up to 200, or the smaller, speedy, and shallow-draft schooners. Pirates never bothered to build or buy their ships. They stole them—always upgrading the size and speed by stealing bigger, better vessels.

What isn't stressed in books and movies is that pirate ships, like all wooden ships of that era, were dark, damp, and filthy. They reeked with the stench of human waste, garbage, and stale bilge water. Pirates used the bow as a bathroom in good weather—otherwise a pot in a covered box below deck. Their urine was saved in barrels to sponge off the cannons since salt water would corrode them, and scarce fresh water was used only for drinking.

Cows, goats, pigs, and chickens were sometimes taken aboard for future food, adding to the filth and smells. Not surprisingly, the ships were a breeding paradise for flies, maggots, spiders, roaches, and rats.

Meals hardly resembled the food on cruise ships. Without refrigeration, it was rancid butter, sour beer, salt pork, dried beef, hard biscuits, and even harder cheese. On one ship, the cheese was so hard it was carved into buttons.

Pirate food was awful!

Pirates floated crackers in their hot tea until the worms crawled out and could be skimmed off. A favorite dish was Salmagundi, a stew of leftover meat, fish, and turtle marinated in wine and spices. Spoiled food was disguised with hot pepper sauce. Chinese pirates kept and bred their own fresh food—rats and caterpillars they cooked with rice. Then, they relaxed by playing cards and smoking opium.

Of course, there was plenty of drinking. In addition to rum, a favorite cocktail was a mix of raw eggs, sugar, sherry, gin, and beer.

The crew slept jammed side-by-side on the deck or below deck in hammocks. Long before Dramamine, pirates had invented their own sure-fire cure for sea sickness. Swallow a piece of pork fat tied to a string and then pull it back up again. Repeat until cured!

Pirates never bathed while at sea. They typically wore jackets of rugged sailcloth, which were coated with tar residue in battle to deflect sword thrusts. Some believed that if they wore calico cotton pants and never washed them, their pants would become bulletproof.

In contrast, some pirate captains dressed like dandies. When Captain Bartholomew Roberts fought his last battle in 1722, he wore a crimson silk waistcoat, red feather in his hat, and fancy boots.

On shore, pirates scrubbed up and dressed to the nines in their stolen goods. It was all silk and lace, powdered wigs and silver buckled shoes. They specially loved wearing gaudy jewelry— elaborate pearl ear pendants, gold necklaces, and diamond-and-emerald crosses stolen from Catholic ships.

Shore stops were not enough to relieve the boredom of cramped ships and long waits between battles so there were on-board card and dice games, impromptu wrestling and boxing matches. The large ships even had musicians who played during meals.

There really were peg-leg and one-eyed pirates because of the dangers of their profession. Those who survived the amputation were given jobs as cooks or carpenters. In those days, the antibiotics were turpentine and lard.

And there really were parrots on pirate shoulders. It was common for seamen who traveled the tropics to buy a colorful bird. Parrots were especially popular because they were easy to care for on a ship, could be taught to talk, lived long, and could be sold back in New York or London for lots of money.

The idea of a secret map with an X marking the spot was good fiction for Treasure Island readers, but bad fact. That legend probably began with the rumors that Captain Kidd hid gold on Gardiners Island off Long Island, New York. But to this day no one has found anything of value—and many have searched.

In fact, most pirates spent their plunder in drinking, gambling, and girls. It was easy come, easy go.

What hasn't been highlighted in books and movies is that pirates organized and ran a truly democratic society. While absolute autocratic rule reigned on other ships and shore, all pirates had an equal vote on everything from electing captains to making and enforcing rules. The captain served with absolute control only during times of battle. At all other times, he served at the pleasure of his crew so he could be fired at any time for bad leadership, cowardice, cruelty, or failure to capture enough treasure. Respect for authority was earned not forced.

Despite slavery on land, black pirates had the right to vote, bear arms, and got equal share of the booty. Pirates were defiant of all outside authority. The only law they respected was their own. Your religion, your color, your sexual preference made no difference. There were no moral judgments.

Pirates even had their own medical disability program whereby part of the plunder was set aside to compensate men disabled in battle. If you lost your arm, you got 600 pieces of eight or $15,000 in today's money. If you lost both eyes, you got $25,000.

Pirates dressed as they pleased, and everyone ate the same food and drank the same rum. There was no officer's mess, private cabins or privileges.

Contrast that to life aboard a Navy ship where ordinary seamen had no control of the rules and were strictly regulated on everything from dress to when they got time off. The captain was king at all times. Make even a small mistake and you were beaten. As one sailor said, "I'd rather be in jail than on a Navy ship. In jail, I won't drown, and I'd have more room, better food, and far better company."

Flogging was common on Navy ships.

Pirates had to sign and obey Articles of Agreement established by their own crew. Typical rules forbade stealing or fighting among the crew. Weapons had to be kept clean, sharp, and loaded. If two men quarreled or fought on board and would not make-up, the matter was solved on shore by sword or pistol.

So serious was desertion in time of battle that offenders could be flogged or have their noses sliced off. Repeat offenders were put to death by hanging or marooned on a deserted island. Unless you were Robinson Crusoe, marooning usually meant slow death from starvation or exposure.

However, these punishments were no harsher than on Navy ships or on shore. In 17th-century England, a child could be hanged for stealing a loaf of bread.

Captured seamen who resisted pirates were usually shot or hacked to death and thrown overboard. A few weren't lucky enough to die quickly. There were cases of prisoners' feet nailed to the deck or lighted matches tied between fingers and toes so that flesh burnt off to the bone. Walking the plank was popular in fiction not fact. But there were instances of keel hauling and dragging prisoners behind the ship. Each pirate captain had his favorite punishment.

In contrast, captured pirates were punished by governments with beheading, prison or hanging.

Hanging pirates was public entertainment.

In England and America, pirate trials were as well attended and publicized as an O.J. Simpson trial. And the actual public hangings were popular entertainment for the masses. In London, thousands would crowd around the gallows, while the wealthy paid high prices for reserved, front-row seats. In a show of bravado, many pirates went to the gallows wearing their best silk and velvet clothes. Some threw gold coins to the watching mob before they swung off.

Spectators waited with baited breath to hear the "last words" of the prisoner. They cheered pirates who "died well" and booed those who didn't. When one tight-lipped pirate was asked if he had any "last words," he snarled, "Nope, I came here to die, not to give a speech!"

In contrast, the final act of another pirate showed his showman talent by picking the pocket of the hangman before he swung off.

Though many pirates died on the gallows from what was called "rope fever," many more died of drink and disease, shipwrecks and battle wounds. Scurvy killed more pirates—and seamen—than sea battles.

However, on the island of Madagascar, dead pirates still "live and drink." Villagers have a ceremony every year where they dig up their long-dead pirate ancestors, take them to their favorite bar for a glass of rum, and then return them to their graves.

Just as pirates practiced democracy long before it was common on land, pirates were also way ahead of the times when it came to women's lib. Women sailed on pirate ships, commanded them and sank with them. As long as 1,500 years ago, Princess Alwilda of Sweden and her all-female crew terrorized the Baltic Sea.

Anne Bonny was born out of wedlock in Ireland around 1690 and grew up in Charleston, South Carolina where she was bigger and stronger than anyone else her age. In her teens, she ran off with pirate Calico Jack Rackham. She dressed in seaman clothes to disguise her sex, and their partnership flourished as they plundered ship after ship. Her fellow pirates said she fought like a wildcat with pistol and cutlass, all the while screaming like a banshee.

Mary Read, also a runaway who dressed like a man, was captured by none other than Anne Bonny. They became fast friends when they revealed their masquerades.

Female pirate Read revealed her sex to a victim.

Read and Bonny were finally captured in 1720 by the British Navy. When they were sentenced to hang, the judge asked if they had anything to say. They replied, "Milord, we plead our bellies." Both were pregnant so the hanging was postponed since by law you couldn't take the life of an unborn child.

Read died of fever in prison before giving birth. Bonny's wealthy father supposedly spirited her away, but no record of Bonny or her child has ever been found.

A century later in China, a growing number of fishermen were turning to piracy to supplement their income. It was not long before a farsighted entrepreneur organized a pirate confederacy that included 70,000 pirates and 1,000 boats.

The so-called "Admiral" of what became the largest pirate navy on the planet was Cheng Sao. Most remarkable, Cheng was a woman, a former prostitute who had married a pirate leader and at his death took over and dramatically expanded the operation. So more than a century before American or British women could even vote, a Chinese woman was running the most powerful pirate force in the world.

Madame Cheng added to her fortunes by charging a fee for membership in her confederation, so she was also a century ahead of Al Capone in setting up a mafia-style protection racket. She also operated loan banks for members in default, and ran a high court to settle differences among her pirate members.

By 1810, a weak and desperate Chinese government finally enlisted the help of British gunboats. Seeing the handwriting on the wall, Madame Cheng along with 17,000 of her pirates accepted amnesty. She returned to Canton and peacefully ran a gambling house until she died at the age of 69. One historian called her, "the greatest pirate, male or female, in all history."

Madame Cheng may have been the world's most powerful pirate, but she wasn't the most famous—or should I say infamous. That dubious honor belongs to a person born in Bristol, England in 1675. He would grow up tall, big, and strong with thick, coal-black hair and blazing black eyes. He ran away to become a merchantman, then successful privateer, and finally turned to piracy. His name was Edward Teach, but everyone called him Blackbeard.

Though only a pirate for 15 months, he was the most audacious that ever lived, and in that short time looted 100 ships throughout the Caribbean and along the Atlantic coast. His command ship was a captured French vessel 100-feet long that had 40 guns and a crew of 150.

In addition to great strength, cunning, and fighting ability, Blackbeard cultivated such an image of terror that most ships surrendered at the sight of him. Seamen believed they were doomed because he was the devil incarnate. It wasn't so much the stories of how he slit throats or forced captors to eat their ears or because he drank prodigious amounts of rum flavored with gunpowder. It was his physical appearance. At a time when few men wore beards, Blackbeard had a long, bushy beard that covered his face and reached to his chest. It was also braided and decorated with colored ribbons.

Blackbeard was the baddest of the bad.

Before going into battle, Blackbeard stuck lighted fuses under his hat. The effect was terrifying. His face, with its fierce eyes and matted hair, was wreathed in smoke so that he looked like a fiend from hell. A walking arsenal, he carried six pistols, a cutlass, and several daggers.

Everything about Blackbeard was larger than life. At six-feet-four, he was a foot taller than the average pirate. He married 14 times, and he could out drink, outfight, and out cruel everyone else.

On a warm May day in 1718, Blackbeard sailed into Charleston, robbed nine ships of gold and slaves, blockaded the harbor for two weeks, and promised to burn down the city unless he got needed medical supplies. The intimidated Governor promptly sent out thousands of dollars worth of medicines, probably for the treatment of syphilis, and Blackbeard sailed away.

But Blackbeard's reign of terror would come to a bloody end on the morning of November 22, 1718, off Ocracoke Island, North Carolina. As British Navy Lieutenant Robert Maynard with 60 men on two sloops neared the pirates, Blackbeard drained a cup of rum, wished Maynard luck, and fired a broadside that killed or wounded 15 of Maynard's men and crippled his sloop.

Under cover of smoke, the pirates boarded Maynard's ship and the hand-to-hand battle began. First with pistols blazing, then the cut and slash of cutlasses.

Finally, it was Blackbeard against Maynard. Despite five pistol shots and 20 cutlass wounds, Blackbeard kept fighting until the swing of a sword cut off his head. The bloody battle was over in 10 minutes.

Legend has it that when Blackbeard's body was thrown overboard, the headless corpse swam around the ship three times. As a historian, I've always wondered whether a headless Blackbeard swam freestyle or breaststroke. What we do know is that Lt. Maynard sailed away in triumph with Blackbeard's head tied to the bow as bloody proof that the most notorious of all pirates was finally dead.

The history of Blackbeard's skull is as colorful as Blackbeard. It was first put on a tall pole in Hampton, Virginia to show that pirate crime did not pay. It supposedly next appeared at the Raleigh Tavern in Williamsburg, Virginia where the cavity was lined with silver and used as a drinking cup. Later it was allegedly used in secret fraternity rituals at the University of Virginia before ending up at the Peabody-Essex Museum in Salem, Massachusetts, where you can see it today.

Blackbeard, Captain Kidd, and Madame Cheng are now history, but not piracy. Instead of plumed hats and silk waistcoats, they wear baseball hats and T-shirts. Instead of pistols and cutlasses, they use machine guns and rocket launchers. The dress and weapons are different, but today's outlaws of the ocean are still pirates.

Based on land and using high-speed motorboats, pirates are still using the age-old tactics of hit-and-run, surprise and terror. Every year, hundreds of pirate attacks occur in the remote waters of the Philippines, Indonesia, Malaysia, and Africa. There are plenty of targets what with 40,000 ships that now move the world's trade across all seven seas. And like their ancestors, these sea bandits will kill for their booty.

Today's pirates steal whatever the vessels carry: money, oil, drugs, aluminum, electronic equipment, even the ship itself. And pirates get high ransoms for captured crews.

In an ironic twist, high-tech pirates using scuba gear were recently arrested in Venezuela. Their crime—looting old pirate shipwrecks without a permit. The idea of pirates stealing from long-dead pirates would have delighted Blackbeard!

* * * * *

FURTHER READING

Under The Black Flag, David Cordingly, Harcourt Brace, 1995.

The Pirates, Time-Life Books, 1978.

Piracy, Days of Long Ago, Kenneth Mulder, Mulder Press, 1998.

Pirates: Terror on the High Seas, D. Cordingly, Turner Pub., 1996.

The History of Pirates, Angus Konstam, Lyons Press, 1999.

A General History of Pirates, Chas. Johnson, Conway Press, 1998.

Pirates, Philip Steele, Lorenz Pub., 1999.

Pirates; Adventurers of the High Seas, David Marley, London, 1995.

The Spanish Main, Peter Wood, Time-Life Seafarers, 1979.

Pirates (Children's book), John Matthews, Atheneum, 2006.

Women Pirates, U. Klausmann & M. Meinzerin, Black Rose Pub., 1997.

CHAPTER 9

THE COLORFUL HISTORY
OF CRUISE SHIPS

Cruise ships–the ocean's biggest wonders.

If a cruise ship could tell her story, it would be far more than her own. It would be the winds that whispered to her, the waves that rocked her, the whales and dolphins that swam with her, and the millions of passengers that have loved, laughed, and learned as she carried them around the world.

The ocean's biggest wonders are also the largest moving objects ever built by human hands. So big that Bob Hope practiced his golf drives on cruise ships. Johnny Weissmuller, better known as "Tarzan," gave his Tarzan yell before diving off a balcony into a ship pool. Cole Porter composed "Begin the Beguine" on a world cruise.

Notably absent from this chapter is the *Titanic* because her story is so well-known. Two little known bits of trivia: the movie lasted longer than the actual sinking, and despite the claim of some historians, the ship's final dinner did not include iceberg lettuce and upside down cake.

* * * * *

The birth of cruising came on a May morning in 1819 in Savannah, Georgia. Experts called her a "madman's gamble," and said she was as likely to get across the Atlantic as reach the moon. She was the *Savannah*, a 99-foot, steam-powered sailboat with a 90-horsepower engine that drove her twin paddles 5 m.p.h. Her tall smoke stack could swivel to direct the smoke and sparks away from the sails.

Steam-powered boats were common by 1819 and had been carrying passengers along coastal waters and even across the Irish and Baltic seas. But what no steamboat had ever done was cross the Atlantic. And that's what Captain Moses Rogers proposed to do with his *Savannah*.

Thanks to fliers and ads in the newspapers, there was plenty of publicity about his daring, first-ever voyage. But there was a big problem—no one bought a ticket! The idea of sailing on a ship across a huge ocean with a fire blazing inside was just too scary for travelers in those days. Scoffers called her "the steam coffin."

To prove it was safe, Captain Rogers persuaded President James Monroe to make a one-day excursion. The President brought along his Secretary of War and five generals. Despite a safe trip and more publicity, still not one passenger bought a ticket. Even the cotton merchants refused to put their cargo on the ship.

The Savannah was the first cruise ship to cross the Atlantic in 1819.

So on May 22, 1819, the *Savannah* sailed to Europe—empty. The only people who enjoyed the ship's 32 staterooms and two salons with tufted couches and Oriental rugs were the crew.

When the *Savannah* neared the Irish coast on the third week, a British ship saw black smoke spewing from her smokestack and rushed to put out the fire. Others on land flung themselves on the ground in prayer thinking she was a fire-breathing monster signaling the apocalypse.

After reaching Liverpool, the *Savannah* steamed on to Stockholm and St. Petersburg where six passengers signed on for the return trip to America. By this time, Captain Rogers was broke so he sold the *Savannah* to owners who ripped out the engine and used her to carry mail and cargo until she ran aground and sank off New York in 1821.

* * * * *

Early cruise ships offered passengers little more than the bare essentials. Cabins were cramped, food poor, and passageways were dank and dark, lit only by small portholes and oil lamps.

Going to bed required acrobatic skill.

The famed American observer, Alexis de Tocqueville, captured the spirit of early 19th-century cruising when he wrote his mother:

> "You cannot imagine what an unusual life one lives in this great stagecoach called a ship. The necessity of living on top of each other and of looking each other in the eye all the time establishes an informality and freedom unlike anything on land. Here each one carries on in the middle of the crowd as if he were alone. Some read aloud; others play cards; sing; write; drink; laugh; or eat as their fancy suggests...

"Our cabins are so narrow, we must go outside to put on our pants. I know not what part of one's toilet that does not take place in the face of someone."

* * * * *

But cruise ships quickly got bigger, faster, and more luxurious.

Britain's Britannia launched regularly scheduled transatlantic service in 1840.

On July 4, 1840, the *Britannia*, a 207-foot paddle-steamer with sails left England for Boston via Halifax, Nova Scotia—at a steady nine knots. She was carrying 115 passengers plus the Royal Mail, and made history as the first regularly scheduled transatlantic ship. Fourteen days after leaving Liverpool, the *Britannia* safely delivered both mail and passengers including the man behind it all, a Nova Scotian ship owner, Samuel Cunard. So began Cunard Line that would rule the Atlantic for more than 100 years, and outlast all competitors in transatlantic travel. Bostonians were so delighted to be chosen as the port of call that they sent Cunard 1,873 invitations to dinner—enough for him to dine out every night for five years.

Cunard was described as "a bright, tight little man with keen eyes, firm lips, and happy manners, who made both men and things bend to his will." He also had wealth, power, and confidence in his vision of a promising—and profitable—future for steam liners.

For seven years, *Britannia* would carry passengers to and from England including Charles Dickens. Plus, a cow to provide fresh milk, and three cats to catch the rats that loved to eat the leather mailbags and glue on the stamps.

The *Britannia* was noted for her meticulously kept routine. Staterooms were swept every morning at 5:00 a.m. The bar was open at a liberal 6:00 a.m.

When her black funnel got hot, the paint kept melting off. So a clever crewman added buttermilk and ocher to the new paint. Now when the funnel got hot, the paint not only fused to the funnel, but turned orangey red. Ever after, all Cunard ships had orangey red funnels.

* * * * *

In the heyday of the Victorian era, the wealthy loved extravagance, whether it be palace-size homes on land or palace-size ships on the sea. And nowhere was that appetite more lavishly indulged than when England built the *Great Eastern* in 1858.

The *Great Eastern* was the mother of all superliners. Her iron hull was 700-feet long. She was six times bigger than anything ever afloat before. Her twin paddles were as big as Ferris wheels. Her propellers were the size of windmills. She could steam around the world without refueling thanks to being able to store 3,000 tons of coal to feed here 10 boilers and 100 furnaces.

The Great Eastern was the first "floating palace."

Because passengers didn't quite trust steam, she also had sails. There was so much canvas on her six masts that on the rare occasions she used full sail, it took 80 men five hours to put them up.

The *Great Eastern*, like the *Great Western* before her, was the brainchild of Isambard Kingdom Brunel, a pioneering engineer who had built successful railroads across England. He boldly decided to extend his line across the Atlantic—and did.

Brunel, in rumpled trousers spattered with dockyard mud and chewing a cigar, was a hands-on boss. As he roamed his shipyard overseeing everything, he wrote copious memos that he stuffed in his top hat for safekeeping.

Brunel in front of the giant chains that launched his gigantic Great Eastern.

The *Great Eastern* had cabins with bathtubs and hot and cold running water for 4,000 passengers. At that time in England, very few homes on land had bathtubs. The talk of the day was that the cabins had double beds for newlyweds.

Meals matched the finest restaurants on land: green turtle soup, goose in champagne sauce, and salmon with hollandaise.

There was a barbershop, spittoons in the shape of seashells, and a food barn on deck for cows, sheep, pigs, and chickens.

For the first time, large deck spaces were provided for promenading that encouraged fashionable clothes and shipboard romances. Deck sports included potato-and-spoon races, blindfold boxing matches, shuffleboard, and deck tennis.

After dinner were other firsts—musicals and lectures. The Reverend Willits became history's first shipboard lecturer when he talked about "Sunshine, The Secret to Happiness."

But bad luck plagued the *Great Eastern* from the start. An explosion delayed builders for a year. On her fourth crossing, high seas tore off the paddle wheels, broke the rudder, and knocked over the cow shed. One of the cows fell through a skylight onto passengers in the lounge. She rolled and pitched so badly that passengers were scared away. Then she caught fire, crashed on a reef, had a mutiny, and bankrupted an owner who died of a stroke.

The *Great Eastern* ended her tragic career as a workhorse laying the first telegraphic cable across the North Atlantic. She was the only vessel big enough to store all 3,000 miles of the cable.

Superstitious sailors said they knew why the ship was jinxed. When her bottom was opened for scrapping, two skeletons were discovered. They were workers who had unknowingly been welded inside the double hull during construction.

If the *Great Eastern* bankrupted her owners, it was mostly because she was a luxury liner whose time had not yet come. It would be 50 years before anyone dared build anything bigger or more luxurious.

* * * * *

But smaller ships filled in, and some pioneered a new concept—cruising to exotic ports of the world far from the transatlantic highway. In 1865, the *Quaker City* took 70 pilgrims on a cruise to the Holy Land.

One not-so-holy passenger would write, "I was looking forward eagerly to months of luxury on the breezy Atlantic and sunny Mediterranean where I could read, watch whales, drink, dance, play cards, promenade, smoke, sing, star gaze, and make love."

That passenger was Mark Twain, and his cruise would be immortalized in his popular book, *The Innocents Abroad*.

Mark Twain was a pioneer cruiser.

Early cruise ships did more than just pleasure passengers. The first commercial use of the newly invented electric lights

was on the steamship *Columbia* in 1880. They were installed by
the inventor himself—Thomas Edison.

* * * * *

Starting in the 1840's, a growing flood of European
immigrants created a demand for cheap transatlantic passage. So
shipowners built bigger and bigger ships. By 1905, 12,000
immigrants were arriving in New York Harbor every day. The
cost for a windowless bunk in a crowded dormitory on the
lowest deck was $50. That compared to $4,000 for a luxury suite
on the top deck of the same ship.

By 1913, more than 40 passenger lines had brought a
million immigrants to Ellis Island. And by the 1930's when the
immigration flood ended, 17 million Europeans had crossed the
Atlantic to begin a new life in the New World.

* * * * *

But transatlantic ships were famous in the eyes of an
admiring public because of the 75% of the liner's space reserved
for the rich and the royal.

The great ocean liners that crisscrossed the Atlantic from the
late 19th-century to the first half of the 20th-century flew the
colors of more than 20 lines of a dozen countries. Their names
alone evoked epic grandeur: the *Leviathan*, the *Normandie*, and
the *Queens, Mary and Elizabeth*. Passage was like a week-long stay
in a luxury resort. As one brochure put it, "A passenger might
now have the privilege of seeing nothing at all that has to do
with a ship—not even the sea!"

Ships of the Gilded Age evoked ornate palaces. Cabin suites
had silk-walled living rooms, fireplaces, and bathrooms with
silver-plated toilet seats. Some passengers rarely left their suites,

preferring to invite friends in for bridge, cocktails, and fancy dinner parties.

Cruise ships in the Gilded Age resembled Baroque birthday cakes.

Even the passengers' dogs enjoyed VIP service with their own menu offering everything from beef consommé to hearts of veal and a vegetarian plate. One ship had different doggie menus for every day of the transatlantic crossing.

* * * * *

No event better epitomized Germany's determination to be a power player on the sea as well as the land than the 1914 launching of the *Vaterland* ("Fatherland"). At 950-feet, she was longer than the tallest building then standing, New York's skyscraping Woolworth Building.

To serve her more than 4,000 passengers was a crew of 1,200 including 60 cooks in eight kitchens. Supplies for just one Atlantic crossing were mind-boggling: 14,000 table napkins, 45,000 pounds of fresh meat, 18,000 bottles of champagne, wine, and brandy, and, of course, premium German beer—all 8,000 gallons of it.

Men smoked cigars in front of massive marble, log-burning fireplaces in rooms with walls of tooled leather, elaborate wood carvings, and stained glass windows.

Shuffleboard was a fancy dress affair on many pre-World War I ships.

But shortly after the *Vaterland's* maiden voyage, the party was over. World War I changed everything. She and 34 other German ships in American ports were seized and converted to Allied troop transports and hospital ships. Renamed the *Leviathan,* she continued to set records by carrying on one crossing 14,000 troops—more humans than ever before on a single ship.

To try to fool prowling German submarines, the *Leviathan* and other ships were painted in zigzag patterns designed to breakup a ship's silhouette so the enemy couldn't tell whether they were coming or going. There was no way to judge its effectiveness, but it was a great morale booster.

A wound to this sailor's mouth helped make him a major movie star.

One American sailor on the *Leviathan* was standing so close to her guns during a test firing that a shell fragment ricocheted into his mouth. The wound caused a minor but permanent change in the way he talked. But the resulting tight-set lip ended up helping him get "tough guy" roles in movies. His name was Humphrey Bogart.

* * * * *

The Roaring 20's fueled by American tourists would set the stage for European nations to compete for passengers. Each tried to outdo the other in building the biggest, fastest, and most luxurious ocean liner. England and France led the way, but later would come Italy, Holland, Norway, and again Germany before World War II.

Passengers were treated like royalty, and dressed the part. New York City Mayor, Jimmy Walker, brought 44 suits, 20 vests, six topcoats, 100 cravats, 30 shirts, and 16 pairs of shoes. One passenger advised that "20 pieces of luggage were an absolute basic minimum for social survival. A traveler should count on at least four changes of clothes per day."

Known as "the ship of light" and hailed as the most beautiful ship ever built, France's *Normandie* was the biggest, fastest, and most sophisticated vessel afloat. She combined the splendor of Versailles with the grace of a yacht. The whole ship smelled of Chanel Number Five!

First-class passengers were greeted by doormen in powdered wigs at the dining room that was longer than the Hall of Mirrors at Versailles. It stretched the length of a football field and was four decks high making it the largest room afloat. The crystal chandeliers were created by famed jeweler and glass designer, Lalique.

To keep passengers busy, there was a full-size swimming pool and tennis court plus evening concerts and ballet performances.

There was even a marble-walled chapel, shooting gallery, and medical center with operating room. At the beauty parlor, hair was shampooed with the best water you could get—distilled water from New York.

France's Normandie had speed and splendor.

How the French loved their floating palace. Exclaimed one exuberant patriot, "Just as our soaring cathedrals embody the Middle Ages, our castles the Renaissance, and our palaces the age of kings, so our *Normandie* embodies all the best of French beauty, grace, and style in the Modern Era."

Sadly, tragedy would strike the *Normandie*. In 1939 after only four years in service, World War II descended like a poisonous fog. Like the *Leviathan* before her, the *Normandie* was painted battle gray, gutted of her treasures, and readied to carry 14,000 soldiers at a time.

But it would never be. Sparks from a welder's torch started a fire that would sink her in the mud of her Hudson River pier. Wrote the New York Times, "The sight of her hurts the human eye and heart." A Wall Street investor bought the $60 million ruined hulk for $160,000, and cut her up for scrap.

* * * * *

Just one year after the *Normandie* went into service, so did Cunard's legendary *Queen Mary*. Following a long tradition of giving their ships names that ended in IA, Cunard officials decided to call the ship, *Victoria*, after their deceased Queen. They went to Buckingham Palace and told King George V, "Your Majesty, we are pleased to inform you that Cunard wishes your approval to name our newest and greatest liner after England's greatest queen. Without a moment's hesitation, the King replied, "My wife, Mary, would be delighted." And that was that!

Now the British had one-upped the French by launching a ship that was bigger and faster than the *Normandie*. The *Queen Mary's* weight of 81,000 tons was more than the entire Spanish Armada put together. She was taller than the Statue of Liberty and Eiffel Tower. All three of Columbus' ships would fit in her restaurant—with room for the *Britannia*, Cunard's first passenger liner.

England's Queen Mary became the world's biggest, fastest ship in 1936.

Someone calculated that the engines that drove the *Queen Mary* across the Atlantic at a record 36 m.p.h. had the muscle power of seven million men rowing in unison. Her rudder was as big as a house. Her 20-foot wide propellers weighed 35-tons each, and were so delicately balanced they could be turned with a finger. She used a gallon of fuel every 13 feet. She would be the world's fastest liner until 1952.

She also held another speed record that would never be broken. British Olympic runner, Lord Burghley, ran one lap around the Promenade Deck in under a minute—in his evening clothes.

I was walking not running on that same Promenade Deck when commuting to England as an Oxford student in the 1950's. One chilly, winter crossing, I was the only passenger walking laps except for a perky little girl with big, brown eyes who asked if she could join me. After four days, we were fast friends. When we ended our hike on the last day, she took me back to meet her parents who were bundled up in deck chairs. My jaw dropped when I realized that her mother was song-and-screen legend, Judy Garland. My young walking partner with the big brown eyes was Liza Minnelli.

Like the *Normandie*, the *Queen Mary* was a floating museum of art and furnishings. Everything was specially designed and fabricated from the door knobs to six miles of inch-thick carpet.

The main lounge featured an etched glass dome, marble fireplaces, and sculpted, deep-pile Wilton carpets. The two-level cocktail lounge blazed with fire-engine red pillars and lamps trimmed in nickel and chrome. A colorful mural of London scenes stretched across the back of the ebony-fronted bar.

In addition to the art-filled interiors, there were 21 elevators, 700 clocks, and enough electricity generated to light up a city of 100,000.

The dog kennels had their own walkways—even lampposts for the British dogs and fire hydrants for the American dogs.

One reporter wrote, "Everybody on the *Queen Mary* is socially prominent, even the dogs!"

Queen Mary's cabin class restaurant was splendiferous.

When the *Queen Mary* arrived—or departed—her whistles could be heard for 10 miles. Wrote E.B. White, "I heard the *Queen Mary* blow one midnight, and the sound carried the whole history of departure, longing, and loss." Those original whistles are still being heard on the *Queen Mary* II.

* * * * *

In 1940, Queen Elizabeth christened her namesake and sister ship to the *Queen Mary*. Called "the noblest vessel ever built in Britain," the *Queen Elizabeth* was a fifth-of-a-mile long (1,031- feet) and 12-feet longer than the *Queen Mary*.

Over the years, *Queen Elizabeth* passengers included the Duke and Duchess of Windsor who boarded with 150 monogrammed suitcases and their pug dog. Tom Mix also came with his horse, and a lady brought her Rolls Royce and six hens so she could be assured of fresh eggs. Cary Grant fell in love while aboard, and Fred Astaire danced in the lounge.

Tugboats take the Queen Elizabeth to sea.

Not surprisingly for a nation famous for its hearty breakfasts, Cunard ships offered eight kinds of bacon, 20 kinds of cereal, 18 different breads, and 15 different jams and marmalades.

* * * * *

But once again, the floating palaces became floating barracks with the outbreak of World War II. Like the *Leviathan* and *Normandie* before them, the *Queens* donned battle garb. The two *Queens* carried two million soldiers from one combat zone to another. Because each could carry 15,000 troops at one time and shift armies from one battlefield to another in days, they played a crucial role in winning the war. They were so important that Hitler offered $250,000 to any submarine commander who sank one. No one collected. Winston Churchill said the *Queens* shortened the war by a year.

Churchill himself crossed the Atlantic three times on the *Queen Mary* during the war. On one trip, he planned the invasion of Normandy by floating toy ships in his stateroom bathtub. Apparently, he spent long hours working in that bathtub since ship carpenters built a special desk that fit across his tub. Cunard also provided a bedside candle in a saucer of water that burned all night—not for light but as a safe way for Churchill to light up his middle-of-the-night cigars.

After the war, the *Queens* helped bring over 13,000 war brides before returning to passenger service during the last boom of transatlantic ship travel.

* * * * *

The 1950s marked the final decade of prosperity for most of the port-to-port ocean liners. A vast, new American middle class wanted to see Europe—from veterans and grandchildren of immigrants to newlyweds and college kids.

I was one of those college kids who crossed the Atlantic several times on World War II vessels converted to one-class student ships. All 800 of us slept in four-berth cabins or 30-bed dormitories. Food was barracks-basic. During the day there were lectures on European history and art and language classes. A jazz band played nightly in a cocktail lounge where Heineken beer was a dime, gin-and-tonics a quarter, and the round-trip fare was $300. Best of all, the girls outnumbered the boys three-to-one!

At our first lunch, I watched my roommate's face turn the color of his green pea soup, clutch his mouth, and run to our cabin violently seasick. Seasickness was common then as ships had no stabilizers. But this was still amazing because our ship hadn't even left the pier.

* * * * *

With a million passengers, mostly Americans, crossing the Atlantic every year, America decided to get in the action with the 1952 launching of the *United States*. Remembering the role of passenger ships in past wars, the *United States* was designed to be converted to a fast troop ship.

The United States was the speediest ever.

To reduce weight, aluminum was used wherever possible— from funnels and deck chairs to lifeboats and oars. Even the flower vases and coat hangars were made of aluminum. The designer was so weight-conscious that he tried to persuade Steinway to build an aluminum piano. But Steinway refused. The obsession with using light-weight aluminum was costly. Infant high chairs in wood cost $20, but the *United States* had them custom-made in aluminum for $250.

By eliminating 2,500 tons from the topside, the *United States* had unprecedented stability and speed. Her massive engines could push her an incredible 44 m.p.h. She would be the fastest liner the world would ever see, able to cross the Atlantic in three days and 19 hours.

That speed record would never be broken, but ironically it was because of planes not faster ships. While the *United States* could cross the Atlantic in a few days, jet planes starting in 1958 did it in a few hours. By 1965, jets were carrying 95% of all transatlantic passengers.

Rumors went around that things were so bad that Cunard could only afford to paint their liners on one side—the side the passengers saw when boarding. After losing millions, Cunard sold both *Queens* in the late 1960s. The *Queen Mary* remains as a floating hotel and tourist attraction in Long Beach, California. The *Queen Elizabeth* was destroyed by fire in the Hong Kong harbor in the 1970s.

In 1969, the *United States* was mothballed, first in Newport News, Virginia and since 1996 in Philadelphia. Her sad claim to fame is that she's the largest abandoned vehicle in America.

* * * * *

But if the golden age of cruise ships was the first half of the 20th-century, the last half and beyond should be called the diamond age. A record 15-million passengers a year now cruise the world basking in luxuries unheard of in the past and in supersized ships five times the size of the *Titanic* that hold up to 8,600 passengers and crew. These floating towns offer dozens of different restaurants and bars, shopping malls, tree-shaded parks, ice skating rinks, mountain climbing, and surfing pools.

Over the past two centuries, cruise ships have ruled the seas—and ruled the hearts of millions. But despite all the changes over the years, the most important aspects of cruising will never change—the magic of a ship and the mystery of the sea. The splash of lines unhooked. The mighty blasts of the whistle as your ship moves away from dockside. A tooting tug pushing the bow as the land gives way to the open sea.

As one traveler put it, "I have never yet begun a cruise that I didn't swear to myself that I would do two things: I would rest, and I would work. I have never done either."

The start or end of a cruiser's day.

* * * * *

FURTHER READING

The Great Liners, Time-Life Books, 1978.

The Only Way to Cross, John Maxtone-Graham, Barnes & Noble, 1972.

The Sway of the Grand Saloon, John M. Brinnin, Delacorte, 1971.

Fifty Famous Liners, F. Braynard & W. Miller, Patrick Stephens, 1990.

Classic Ocean Liners, F. Braynard, Patrick Stephens Ltd., 1990.

The Cunard Story, Howard Johnson, Whittet Books, 1987.

Queen Mary: Her Early Years Recalled, C.W.R. Winter, Patrick Stephens, Ltd., 1986.

Seven Centuries of Sea Travel, from the Crusaders to the Cruises, B.W. Bathe, Leon Amiel Pub., 1973.

Made in the USA
Charleston, SC
17 April 2011